Contents

Introduction

■ ■ ■

Content Guidance

■ ■ ■

Questions and Answers

Introduction

About this guide

This guide is one of the series covering the Edexcel specification for AS and A2 chemistry. It offers advice for effective preparation for the Unit 3 Laboratory Skills internal assessment and the Unit 6 Laboratory Skills internal assessment.

Four distinct skills are required in advanced level practical chemistry.

A General practical competence

Each student will have to perform a number of experiments during each year. There are no marks for this activity. All that has to be recorded is the date and title of the practical. As a student, you will not know when you have done one of these tasks. There are no marks for it — and there is no pass or fail either. All that the exam board requires is that you do a minimum of five of the listed experiments across the three areas of physical, inorganic and organic chemistry at AS and another five at A2. There is nothing you can do in preparation for these experiments that will help you to get a better grade in the unit.

B Qualitative observation

This is marked out of 14.

There are four experiments for AS and four for A2. These have been designed by the Edexcel examining team. You will be issued with a worksheet at the start of the practical session and will have 1 hour in which to finish the practical work. You will be allowed the *Edexcel data booklet*, if it is necessary, but no notes or other books, including this one, will be allowed into the laboratory. All work must be handed in at the end of the session.

The unknown substances will be changed from year to year, but there is little scope for change — especially in organic chemistry at AS.

The focus of the tasks is your skill at making sensible observations and most of the marks will be awarded for this. You will be expected to identify the unknowns, and there will be one or two questions asking for equations or types of reaction.

Only one experiment will count but you are allowed to do two, three or all four — only the best will score. You are not allowed to repeat any experiment.

This guide will help you to prepare for the assessment. Your teacher will warn you in advance that you will be doing, for instance, an AS inorganic qualitative observations assessment. You must prepare for the assessment by reading through the appropriate part of the AS or A2 Activity B section.

Each section has a list of tests and the observations and deductions that you should make from those tests. The ✓ beside the observation informs you of the type of observation that will be expected. There is also a practice test that you can do. If possible this should be done in the laboratory, but if that is not possible then look at the identity of the unknown and write down the observations that you would expect to make. The answers are at the back of the book.

In the Unit 3 and Unit 6 sections there are two mock Activity B practical experiments that you can try, so long as they are supervised and have been risk-assessed by a chemistry teacher. The answers to these are also at the back of the book.

C Quantitative measurement

This is marked out of 14.

AS

There are two titrations and two enthalpy change experiments. As with Activity B, you may do more than one — only the best will count.

The practical work must be completed in a 1 hour session, but you may be allowed to do the calculation at a later time. If this is so, make sure that you know how to do this type of calculation before the second session.

Titrations

Over half of the marks are awarded for choosing two titres that are concordant and for getting the correct mean titre.

The titration will either be acid/base or iodine/thiosulfate. In both you should expect to weigh out a solid, dissolve it in water, transfer it to a volumetric flask and make the solution up to $250 \, cm^3$.

Enthalpy change experiments

The procedure is similar in both experiments. The reagents are mixed in a polystyrene cup that is held in a beaker, and the temperature change is measured. This could involve either adding the second reagent in portions or adding it all at the same time. In the former, a graph of temperature against time will have to be plotted.

The majority of the marks are for the accuracy of measuring the temperatures and for doing the $\Delta H_{reaction}$ calculation.

A2

The four experiments at this level are two titrations and two kinetic experiments.

As with the AS Activity C experiments, the practical work must be completed in a 1 hour session, but you may be allowed to do the calculation at a later time. If this is so, make sure that you know how to do this type of calculation before the second session.

Titrations

The two experiments are:

(1) *A titration involving potassium manganate(VII)*

As with the AS titrations, a solid has to be weighed out, dissolved, transferred and made up to 250 cm³. The potassium manganate(VII) solution will be in the burette and no indicator is needed because no more is added when the solution turns pale pink. Over half of the marks will be for accuracy.

(2) *A pH titration*

Here a strong alkali is added to a weak acid in small portions and the pH of the solution is measured after each addition.

Kinetic experiments

One of these will involve following a reaction, such as the iodination of propanone. The other is to measure the rate of reaction at different temperatures, and so be able to calculate the activation energy.

D Preparation

This is marked out of 12. Being able to calculate the percentage yield gains 3 marks, with 1 or 2 marks available for obtaining a good yield in the experiment.

AS

There are three experiments — one is an organic preparation and the other two involve the preparation of inorganic salts.

A2

There are three experiments — two are organic preparations and the other is the preparation of a solid complex transition metal compound.

> You will **not** be allowed to take this or any book (other than a data booklet) or your notes into the laboratory when you are being assessed.
>
> You are **not** allowed to take out any work from the laboratory, even if the assessment is not finished. Your work will be given back to you when you continue the assessment at a later time.

Content
Guidance

This section is a guide to what you may be required to do in the internally assessed practicals in both Units 3 and 6. It is not a laboratory guide and all health and safety measures should be taken by a teacher before any experiments are carried out by a student.

The topics for both Units 3 and 6 are:
- Activity A — General practical competence (GPC)
- Activity B — Qualitative observations on both organic and inorganic compounds
- Activity C — Quantitative measurement
- Activity D — Preparation of organic and inorganic compounds

Examiner's comments

Many of the observations or inferences are followed by examiner comments which are preceded by the icon 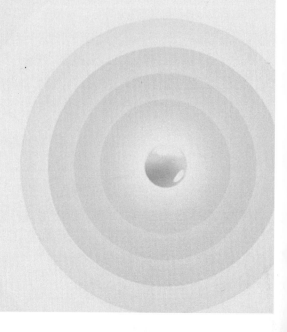. These comments may explain the correct answer or point out common errors.

Unit 3

Activity B: Qualitative observation

There are four experiments that you could be asked to carry out.

(1) An observation exercise on three inorganic unknowns.

(2) An observation exercise on another three inorganic unknowns.

(3) An observation exercise on three organic compounds, probably limited to alkenes and alcohols.

(4) An observation exercise on two organic compounds, one probably being a halogenoalkane.

If you do more than one, each will be marked and you will score the one with the highest marks. Each experiment has a maximum of 14 marks.

Most of the marks are awarded for making the correct observations, so make sure that you:

- mix the reagents properly before recording colour changes
- state if there is a precipitate and record its colour
- make sure, if you are asked to add a substance until in excess, that you observe any changes after adding a little as well as those after adding excess
- state the colour before and after if there is a colour change — if there is no colour change, say so (rather than say that there is no reaction)
- in organic chemistry, if two layers form, say so

Inorganic substances

Most of the marks are for observations. You will normally be expected to identify each solid and you may be required to write an equation, e.g. for a thermal decomposition, a redox reaction or a precipitation reaction. A typical mark distribution is:

Observations	9 or 10 marks
Identification of each unknown	3 marks
Equation and/or comment	1 or 2 marks

⚡ If you are asked to *identify* a substance, give either its name or its formula. If you give both then both must be correct. Names are easier than formulae, where mistakes such as $NaCO_3$ or $MgNO_3$ are common. However, if you are asked for the formula, the name will not score.

You may assume that only the following ions will be tested for.

Cations	Li^+	Na^+	K^+	Mg^{2+}	Ca^{2+}	Sr^{2+}	Ba^{2+}	NH_4^+
Anions	Cl^-	Br^-	I^-	NO_3^-	CO_3^{2-}	HCO_3^-	SO_4^{2-}	

Tests on solids

Heating the solid

Hydrogencarbonates, such as sodium hydrogencarbonate ($NaHCO_3$), and salts containing water of crystallisation give off water vapour, which will condense in the upper part of the test tube. For these, there will be a mark for observing the droplets of water near the mouth of the test tube.

Heating the solid is also a test which indicates if the unknown is a nitrate, a carbonate or ammonium chloride.

The solid should be heated, gently at first and then strongly. Look for sublimation or melting and whether a gas is evolved or not. You might be told to test the gas. Possible observations marks are shown with a ✓.

Test	Observations	Inferences
Heating a nitrate — test any gas evolved with a glowing splint	• Melts ✓ and then gives off bubbles of a colourless gas which relights ✓ a glowing splint	• It is either sodium nitrate or potassium nitrate
	• Melts ✓ and then gives off a brown gas ✓ and one which relights ✓ a glowing splint Note: If the solid is a hydrated nitrate, water/steam will also be observed and will score a mark	• It is either lithium nitrate or a group 2 nitrate
Heating a carbonate or hydrogencarbonate — test any gas evolved with limewater	• Drops of water condense on upper part of the tube ✓ and a gas is produced which turns limewater milky ✓ (or cloudy or gives a white precipitate)	• It is a group 1 hydrogencarbonate
	• (No water produced but) a gas is evolved which turns limewater milky ✓	• The solid is a group 2 carbonate or lithium carbonate
Heating ammonium chloride	• Solid sublimes ✓ (or solid reforms in upper part of the test tube)	• It is ammonium chloride

The equation for the thermal decomposition of potassium nitrate is:

$$2KNO_3(s) \rightarrow 2KNO_2(s) + O_2(g)$$

The equation for the thermal decomposition of magnesium nitrate is:

$$2Mg(NO_3)_2(s) \rightarrow 2MgO(s) + 4NO_2(g) + O_2(g)$$

Flame test

Make sure that the wire is clean by dipping it into some concentrated hydrochloric acid and then putting it into the hot part of a Bunsen flame. Repeat until the flame is not coloured.

Now dip the wire into concentrated hydrochloric acid and then into the solid, and immediately place it in the hot part of the Bunsen flame.

Colour	Ion responsible
Crimson*	Li^+
Yellow (not orange)	Na^+
Lilac	K^+
Yellow–red*	Ca^{2+}
Red*	Sr^{2+}
Pale green (or apple green)	Ba^{2+}

*You must be able to distinguish between these three red flames so as to be able to identify the ion, but to name the actual colour you can give various shades of red — such as magenta for lithium and brick-red for calcium.

Addition of acid: test for carbonates and hydrogencarbonates

When dilute hydrochloric or sulfuric acid is added to the solid, a gas may be evolved.

The identity of the cation may already be known from a flame test. Solid group 2 hydrogencarbonates do not exist.

Test	Observations	Inferences
Acid + solid group 1 compound — test any gas evolved with limewater	• Bubbles ✓ (or fizzing or effervescence) • Limewater turns milky/cloudy ✓	• It is a carbonate or hydrogencarbonate
Acid + solid group 2 compound — test any gas evolved with limewater	• Bubbles ✓ (or fizzing or effervescence) • Limewater turns milky/cloudy ✓	• It is a carbonate

Typical equations are:

$$Na_2CO_3(s) + H_2SO_4(aq) \rightarrow Na_2SO_4(aq) + H_2O(l) + CO_2(g)$$

$$CaCO_3(s) + 2HCl(g) \rightarrow CaCl_2(aq) + H_2O(l) + CO_2(g)$$

To distinguish between a carbonate and hydrogencarbonate

Test	Observations	Inferences
Add some solid to almost boiling water — test any gas evolved with limewater	• Bubbles (effervescence) ✓ • Limewater goes milky ✓	 • It is a hydrogencarbonate
Make a solution of the unknown and add calcium chloride solution	• Either white precipitate ✓ • Or no precipitate ✓	• It is a carbonate • It is a hydrogencarbonate

Addition of sodium hydroxide: test for ammonium salts

When dilute sodium hydroxide is added to the solid and the test tube gently warmed, a gas may be evolved which can be tested with damp red and damp blue litmus paper, or with a glass rod dipped into concentrated hydrochloric acid.

Test	Observations	Inferences
Warm with dilute sodium hydroxide and test any gas evolved: • either with damp red and blue litmus paper • or with a glass rod dipped into concentrated hydrochloric acid	• Gas turns red litmus blue ✓ Blue litmus stays blue ✓ • White smoke ✓	• Gas is ammonia, so it is an ammonium salt • White smoke is ammonium chloride, so it is an ammonium salt

The equation for the reaction of ammonia gas with hydrochloric acid is:

$$NH_3(g) + HCl(g) \rightarrow NH_4Cl(s)$$

Concentrated sulfuric acid test for inorganic halides

This test is used to distinguish between different halides. The concentrated acid is carefully added to the solid. **This test must be done in a fume cupboard**.

Test	Observations	Inferences
Add the acid to a chloride and test any gas evolved: • either with a glass rod dipped in concentrated ammonia • or with damp red and with damp blue litmus paper	• Bubbles ✓, steamy (or misty) fumes ✓ evolved (do not say white smoke here) • White smoke ✓ • Red litmus stays red ✓ Blue litmus goes red ✓	• Hydrogen chloride evolved • It is a chloride
Add the acid to a bromide	• Steamy fumes ✓ and red–brown gas ✓	• Hydrogen bromide evolved • Bromine evolved • It is a bromide
Add the acid to an iodide	Any two of: • Steamy fumes • Purple vapour • Yellow solid • Bad egg smelling gas	• Hydrogen iodide, HI • Iodine • Sulfur • Hydrogen sulfide, H_2S • It is an iodide

Tests on solutions

Addition of acidified silver nitrate solution: the halide test

A few drops of dilute nitric acid are added to the unknown solution, followed by a few drops of silver nitrate solution. The solubility of the precipitate formed is then tested in dilute and concentrated ammonia.

Test	Observations	Inferences
Add dilute nitric acid to a solution of the unknown and then silver nitrate* Test the precipitate: first with dilute ammonia and then (if no change) with concentrated ammonia	• White precipitate ✓ turning purple on standing ✓; soluble in dilute ammonia ✓ • Cream precipitate ✓ insoluble in dilute but soluble in concentrated ammonia ✓ • Yellow precipitate ✓ insoluble in concentrated ammonia ✓	• It is a chloride • It is a bromide • It is an iodide

* Sometimes the silver nitrate is added first, followed by the nitric acid. The observations for a chloride then are white precipitate ✓ which stays on addition of nitric acid ✓.

The ionic equation for the precipitation of a silver halide (X stands for Cl, Br or I) is:

$$Ag^+(aq) + X^-(aq) \rightarrow AgX(s)$$

Addition of chlorine water

This is a redox reaction in which the chlorine oxidises the unknown.

Test	Observations	Inferences
Add a few drops of chlorine water	• Brown (or yellow) solution ✓	• Bromine or iodine produced
Then add starch solution	• Goes blue–black ✓ • Does not go blue–black ✓	• It is an iodide • It is a bromide

The ionic half-equation for the reduction of chlorine is:

$$Cl_2(aq) + 2e^- \rightarrow 2Cl^-(aq)$$

Addition of acidified barium chloride: the sulfate test

A few drops of dilute hydrochloric acid are added to the unknown solution followed by a few drops of barium chloride solution.

Observations	Inferences
White precipitate ✓*	It is a sulfate

* Sometimes the barium chloride is added first, followed by the hydrochloric acid. The observations then are white precipitate ✓ which stays on addition of hydrochloric acid ✓.

The ionic equation for the precipitation of barium sulfate is:

$$Ba^{2+}(aq) + SO_4^{2-}(aq) \rightarrow BaSO_4(s)$$

Test for nitrates

In this test the nitrate ions are reduced to ammonia gas.

Test	Observations	Inferences
Add Devarda's alloy (or aluminium powder) and dilute sodium hydroxide and warm — test any gas evolved: • either with a glass rod dipped in concentrated hydrochloric acid; • or with damp red litmus	• Bubbles (effervescence) ✓ • White smoke ✓ • Litmus goes blue ✓	 • It is a nitrate

Test the solubility of hydroxides

All group 1 hydroxides and barium hydroxide are soluble in water. Magnesium hydroxide is insoluble and calcium hydroxide slightly soluble.

Test	Observations	Inferences
Add dilute sodium hydroxide to a solution of the unknown	• White precipitate ✓ • Slight white precipitate ✓	• It is a magnesium compound • It is a calcium compound

The ionic equation for the precipitation of magnesium hydroxide, for example, is:

$$Mg^{2+}(aq) + 2OH^-(aq) \rightarrow Mg(OH)_2(s)$$

Addition of zinc or magnesium

If the unknown is an acid, such as dilute sulfuric acid, hydrogen will be produced.

Test	Observations	Inferences
Add zinc or magnesium to a solution of the unknown and warm Test any gas evolved with a burning splint	• Bubbles (effervescence) ✓ • Ignites with a squeaky pop ✓	 • It is an acid

The ionic equation for the reaction of magnesium with the acid is:

$$Mg(s) + 2H^+(aq) \rightarrow Mg^{2+}(aq) + H_2(g)$$

Practice example ASB1

Assume that you have been given some solid potassium hydrogencarbonate.

Write down the observations that you would expect to make if you carried out the following tests. The answers are on p. 85.

(a) Carry out a flame test on some of the solid. Write down your observations. (1)

(b) Put some of the solid in a test tube and heat it in a Bunsen flame. Test the gas evolved with limewater. Write down your observations. (2)

(c) Make a solution with the remainder of the solid and add 5 drops of calcium chloride solution. Write down your observations. (1)

(d) Using your observations, write the formula of the unknown solid. (1)

Organic substances

The only types of compounds examined in the AS chemistry course are:

- hydrocarbons — such as hexane and cyclohexene
- alcohols — including primary, secondary and tertiary alcohols
- halogenoalkanes — including chloro-, bromo- and iodoalkanes.

If you do an organic assessment *before* you have studied halogenoalkanes, then the unknowns are limited to hydrocarbons and alcohols.

Always read the question carefully. It might state the number of carbon atoms in molecules of the unknowns, or that two of the compounds are isomers, or that one contains a branched carbon chain.

A typical mark distribution is:

Observations	9 or 10 marks
Interpreting mass or IR spectra	1 or 2 marks
Interpreting observations	1 or 2 marks
Identifying unknowns	2 or 3 marks

Possible observations marks are shown with a ✓.

Combustion

The unknown is placed on a crucible lid and ignited with a lighted splint.

Observations	Inferences
It burns with a clear non-smoky flame ✓	Low carbon to hydrogen ratio — could be an alcohol
It burns with a smoky flame ✓	High carbon to hydrogen ratio — probably a cyclic alkane or cyclic alkene

Addition of water, followed by universal indicator solution or pH paper

The unknown is added to some water in a test tube and shaken.

Test	Observations	Inferences
Add to water	• Either two layers form ✓ • Or it dissolves fully ✓	• It does not hydrogen-bond with water • It hydrogen-bonds with water (and is an alcohol)
Add a little universal indicator	• Green colour ✓	• It is a neutral substance

Tests for a C=C group in alkenes

Test	Observations	Inferences
Add some bromine water to the unknown — stopper the test tube and shake carefully	Brown bromine water goes colourless ✓ and two layers are formed ✓	C=C group present
Add dilute sulfuric acid then potassium manganate(VII) solution — shake the test tube or warm it in a beaker of hot water	Purple solution goes colourless ✓ and two layers are formed ✓	C=C group present

Tests for an OH group

Test	Observations	Inferences
Add a small piece of sodium to the unknown in an evaporating basin	• Bubbles evolved ✓ • Sodium disappears or a white solid forms ✓	• It contains an OH group*
Add solid phosphorus(V) chloride and test any gas evolved: • either with a glass rod dipped in concentrated ammonia; • or with damp blue litmus paper	• Steamy fumes evolved ✓ • White smoke formed ✓ • Litmus goes red ✓	• It contains an OH group*

*At AS, this test shows that the unknown is an alcohol. Carboxylic acids also give a positive result, but they are covered in A2 not AS.

Oxidation to differentiate between tertiary and other alcohols

Test	Observations	Inferences
Add dilute sulfuric acid and aqueous potassium dichromate(VI) — then warm the mixture in a beaker of hot water	• Either the orange solution turns green ✓ • Or the solution stays orange ✓	• It is a primary or a secondary alcohol* • It is a tertiary alcohol†

*A primary alcohol has one carbon atom attached to the –COH group, e.g. propan-1-ol,

$$H-\overset{\overset{\displaystyle H}{|}}{\underset{\underset{\displaystyle H}{|}}{C}}-\overset{\overset{\displaystyle H}{|}}{\underset{\underset{\displaystyle H}{|}}{C}}-\overset{\overset{\displaystyle H}{|}}{\underset{\underset{\displaystyle H}{|}}{C}}-OH$$
. A secondary alcohol has two carbon atoms attached to the

–COH group, e.g. propan-2-ol, $H-\overset{\overset{\displaystyle H}{|}}{\underset{\underset{\displaystyle H}{|}}{C}}-\overset{\overset{\displaystyle H}{|}}{\underset{\underset{\displaystyle OH}{|}}{C}}-\overset{\overset{\displaystyle H}{|}}{\underset{\underset{\displaystyle H}{|}}{C}}-H$.

†A tertiary alcohol has three carbon atoms attached to the –COH group, e.g.

methylpropan-2-ol, $H_3C-\overset{\overset{\displaystyle CH_3}{|}}{\underset{\underset{\displaystyle OH}{|}}{C}}-CH_3$.

Tests for halogenoalkanes

You may be asked to test the solubility in water — halogenoalkanes are insoluble, and so form two layers.

You may be asked to hydrolyse the halogenoalkane to an alcohol, and then test for the halide ion produced.

You may be asked to carry out an elimination reaction (only with secondary or tertiary halogenoalkanes).

Test	Observations	Inferences
Shake a few drops of the unknown with water	• Two layers form ✓	
Add a few drops of ethanol and aqueous silver nitrate and dilute nitric acid — stand the test tube in a beaker of hot water	• White precipitate ✓ • Cream precipitate ✓ • Yellow precipitate ✓	• It contains a C–Cl group • It contains a C–Br group • It contains a C–I group
Add 4 drops of ethanol and then 2 cm³ of dilute sodium hydroxide — stand the test tube in a beaker of hot water for 5 minutes Then add excess nitric acid followed by aqueous silver nitrate	• White precipitate ✓ • Cream precipitate ✓ • Yellow precipitate ✓	• It contains a C–Cl group • It contains a C–Br group • It contains a C–I group
Add concentrated sodium hydroxide in ethanol — fit the test tube with a delivery tube and warm gently, passing any gas evolved through a little bromine water	• Brown bromine goes colourless ✓	• H–halogen eliminated forming an alkene

The identity of the halogen in the precipitate formed with silver nitrate can be confirmed by adding ammonia solution.

Observations	Inferences
Precipitate dissolves in dilute ammonia ✓	AgCl — the unknown is a chloroalkane
Precipitate insoluble in dilute but soluble in concentrated ammonia ✓	AgBr — the unknown is a bromoalkane
Precipitate insoluble in concentrated ammonia ✓	AgI — the unknown is an iodoalkane

Mass spectra

Look for the following.

Peaks	Inferences
At $m/e = 29$ or at (molecular ion value $- 29$)*	C_2H_5 group in the molecule (at A2, this peak could also be caused by CHO in the molecule)
A strong peak at $(M - 1)$*	Caused by loss of H from an alcohol
Molecular ion doublet of equal intensities at two m/e units apart	Molecule contains one bromine atom
Molecular ion doublet in ratio $3 : 1$ at two m/e units apart	Molecule contains one chlorine atom

*For example with propan-1-ol ($M_r = 60$), a peak at $m/e = (60 - 29) = 31$ and a peak at $(60 - 1) = 59$.

You may be asked to work out the number of carbon atoms, and hence the molecular formula, from the value of *m/e* for the molecular ion. For example, an alcohol that is *not* oxidised by acidified potassium dichromate(VI) has a molecular ion peak of *m/e* = 74. The OH contributes 17, leaving 57. This is made up of 4 carbons atoms (4 × 12 = 48) and 9 hydrogen atoms, so the unknown is:

$$H_3C - \underset{\underset{OH}{|}}{\overset{\overset{CH_3}{|}}{C}} - CH_3$$

Infrared spectra

You will be allowed to use the *Edexcel data booklet* to look up the frequencies at which bonds absorb, so you do not need to learn these values. At AS, this will be limited to the values listed in the following table.

Bond	Absorption frequency/cm^{-1}
C–H in alkanes	3000–2853
C=C in alkenes	1669–1645
O–H in alcohols	A broad peak at 3750–3200

The O−H peak is broad because of hydrogen bonding.

Work out the observations in the practice example below.

Practice example ASB2

Assume that you have been given some liquid 2-methylpropan-2-ol. Write down the observations that you would expect to make. The answers are on p. 85.

Compound X has molecules that contain 4 carbon atoms. Carry out the following tests on compound X.

(a) Put 2 cm^3 of water in a test tube and add and equal volume of X, then add 3 drops of universal indicator solution. Write down your observations. (2)

(b) Working in a fume cupboard, put 1 cm^3 of X in a test tube and add half a spatula full of phosphorus(V) chloride and test any gas evolved with damp blue litmus paper. Write down your observations. (2)

(c) To 1 cm^3 of X in test tube add an equal volume of dilute sulfuric acid, followed by 6 drops of potassium dichromate(VI) solution. Stand the test tube in a beaker of hot water for 5 minutes. Write down your observations. (1)

(d) Look at the spectrum opposite and identify the bonds responsible for the peaks labelled P and Q. (2)

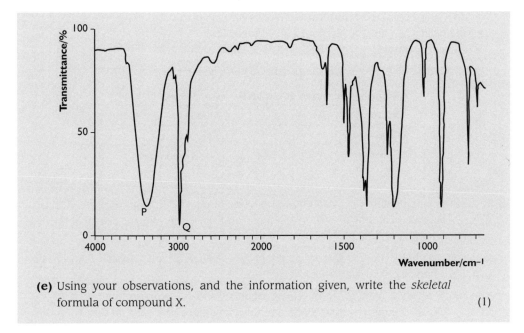

(e) Using your observations, and the information given, write the *skeletal* formula of compound X. (1)

Activity C: Quantitative measurement

Full details of all the experiments assessing this activity will be given to you just before starting the assessment. It is essential that you read the details carefully before starting any practical work.

There are four experiments that you could be asked to carry out:

(1) an acid/base titration

(2) an iodine/sodium thiosulfate titration

(3) an experiment to find the enthalpy change of a single reaction

(4) an experiment to find the enthalpy change for two reactions, and use the results and Hess's law to calculate the enthalpy change for a third reaction.

If you do more than one experiment, each will be marked and you will score the one with the highest marks. Each experiment has a maximum of 14 marks.

Titrations

Over half of the marks are awarded for choosing two titres that are concordant (no more than $0.2\,cm^3$ between the largest and the smallest) and for getting the correct mean titre. This means that the most important skills are:

- weighing out the solid and making up $250\,cm^3$ of solution
- doing the titration and knowing the colour of the indicator at the end point — if you are adding alkali from the burette, stop when phenolphthalein goes just pink or when methyl orange goes orange.

There are two types of titrations that could be set:
- acid/base
- iodine/thiosulfate

For both types the mark distribution is approximately:

Recording of masses and volumes	2 or 3 marks
Accuracy	7 marks
Calculation	3 or 4 marks
Comment	1 or 2 marks

Weighing

You will be supplied with a weighing bottle containing a solid. You must weigh this, using a balance accurate to 0.01 g. The solid is then tipped into a beaker and the weighing bottle reweighed.

You must record the masses and the weight of the solid to 2 decimal places. Thus a mass of 12.30 g must be recorded as 12.30 g and **not** 12.3 g.

The solid will then need to be dissolved in water and the solution poured, using a funnel, into a volumetric flask. The beaker and any glass rod used in dissolving the solid must be washed, with the water going into the volumetric flask. This is then made up to the mark with distilled water and thoroughly shaken. If the shaking is not done properly, inconsistent titres will be obtained and marks will be lost.

You now have two solutions — one made by dissolving the solid, and the other supplied by your teacher.

Errors to avoid in a titration

- *General.* Make sure that you read the instructions carefully before starting the titration. For acid/base titrations, some indicator is added to the conical flask. In iodine/thiosulfate titrations, acid is added initially and the starch indicator must not be added until the iodine colour has faded to a pale yellow (straw) colour.
- *Pipette.* This must be washed out with a little of one solution and then 25.0 cm³ pipetted into a conical flask. Make sure that there are no air bubbles in the stem of the pipette when you are doing this and that you use the correct solution.
- *Burette.* This must be washed out with the other solution. It is then filled to just above the zero mark, and liquid is run out until there is no air in the stem. The funnel must then be removed from the top of the burette before titrating. The initial volume is then recorded.
- *Recording.* You must read the volume by using the level of the bottom of the meniscus as shown.

Make sure that you record the final and initial readings in the correct rows in the table. The initial reading may be 0.00 cm^3, but never 50.00 cm^3.

The titre is the difference between the final and the initial volumes.

- *All volume readings and the titre must be recorded to 0.05 cm^3.* Thus the reading in the diagram on p. 20 is 23.70 cm^3, **not** 23.7 cm^3 and certainly not 24.30 cm^3.
- *Consistency*. You must do repeat titrations until you get at least two that are concordant. This means that the difference between them must be 0.20 cm^3 or less. In Table ASC1, titration numbers 2 and 3 are the concordant titres, and these are the ones that must be averaged to give the average titre that you will use in the calculation.

Table ASC1

Titration number	1	2	3	4
Burette reading (final)/cm^3	24.15	25.05	27.10	25.55
Burette reading initial/cm^3	0	1.20	3.15	2.00
Titre/cm^3	24.15	23.85	23.95	23.55

Use titrations numbers 2 and 3 to calculate the mean:

$$\text{mean (average) titre} = \frac{23.85 + 23.95}{2} = 23.90 \, \text{cm}^3$$

- *Decimal places*. Remember to record all your volumes to 0.05 cm^3. If you had written the mean titre as 23.9 cm^3 or the final reading in titration 3 as 27.1 cm^3, you would have lost a mark.

Marking

- *Weighing*. There is 1 mark for recording all the masses to 0.01 g.
- *Accuracy*. There are 4 marks for getting an accurate mean titre. If your value is ±0.30 cm^3 of your teacher's value, you will score 4 marks. If your value is between ±0.30 and ±0.40 cm^3 away from your teacher's value, you will score 3 marks. If you are more than 0.70 cm^3 from that value, you will score 0 for accuracy.

e If there is no weighing involved, you will need to get within ±0.20 cm^3 of the teacher's titre to score 4 marks.

- *Range of titres*. There are up to 3 marks for how concordant the titres are that you used to calculate the mean. If they were 0.20 cm^3 apart or less, you will score the 3 marks. If the titres in experiments 2 and 3 in Table ASC1 were used to calculate the mean, the range mark would be 3. If titration 4 had also been used th~~ between the largest (23.95) and the smallest (23.55) is 0.40 cm^3 a mark would have been scored. If titration 1 had also been included of 0.90 cm^3 would have resulted in no range marks being awarded. that care must be taken to choose two or more titres with a range of 0.20 cm^3.

Worked example ASC1

Comment on the errors in the following:

Mass of weighing bottle + solid	12.3 g
Mass of weighing bottle empty	11.2 g
Mass of solid	1.1 g

Answer

The masses should have been recorded to 2 decimal places.

Worked example ASC2

Comment on the errors in the following

Titration	1	2	3	4
Burette reading final/cm³	23.45	23.50	45.85	22.25
Burette reading (initial)/cm³	50	1.20	23.50	0
Titre/cm³	26.55	22.3	22.35	22.25

$$\text{Mean titre} = \frac{22.3 + 22.35 + 22.25}{3} = 22.3 \, \text{cm}^3$$

Answer

In titration 1, a full burette has a reading of 0, not 50.

In titration 2, the titre and the mean titre should have been given to 2 decimal places, that is 22.30 not 22.3. It is acceptable for an initial reading of zero to be written as 0 rather than 0.00.

Acid/base titrations

There are two likely types of calculations that you may be required to carry out, but both start in the same way because both have a similar experimental method.

Calculation to find the molar mass of the solid

The method is to weigh out the solid, A, and use this to make up 250 cm³ of solution. This is then titrated against a standard solution of B.

The calculation is best broken down into steps.

(1) Calculate the amount (moles) of substance in the mean titre:

$$\text{number of moles of B} = \text{concentration of B} \times \frac{\text{mean titre}}{1000}$$

(2) Use the stoichiometry to calculate the amount (moles) of the solid dissolved in a 25.0 cm³ sample, and hence the amount in 250 cm³:

$$\text{number of moles of A} = \frac{\text{number of moles of A in equation}}{\text{number of moles of B in equation}} \times \text{moles of B}$$

Hence, the number of moles of the solid in $250\,cm^3 = 10 \times$ above value.

(3) Use your answer to **(2)** and the mass of solid used to calculate the molar mass of the solid:

$$\text{molar mass of A} = \frac{\text{mass of A used}}{\text{answer to (2)}}$$

Worked example ASC3

1.53 g of a dibasic acid, H_2X, was weighed out, dissolved in water and the solution made up to $250\,cm^3$ with distilled water.

$25.0\,cm^3$ portions of this solution were titrated against $0.111\,mol\,dm^3$ sodium hydroxide solution using phenolphthalein indicator. The mean titre was $21.90\,cm^3$.

The equation for the reaction is:

$H_2X(aq) + 2NaOH(aq) \rightarrow Na_2X(aq) + 2H_2O(l)$

What is the molar mass of the acid H_2X? (3)

Answer

Amount of sodium hydroxide in the mean titre $= 0.111\,mol\,dm^{-3} \times \dfrac{21.90}{1000}\,dm^3$

$$= 0.002431\,mol \checkmark$$

Amount of H_2X in $25.0\,cm^3$ of solution $= \frac{1}{2} \times 0.002431\,mol$

$$= 0.001215\,mol$$

The moles of sodium hydroxide are multiplied by $\frac{1}{2}$ to get the moles of H_2X because there is 1 H_2X to 2 NaOH in the equation.

Hence, the amount of H_2X in $250\,cm^3$ of solution $= 10 \times 0.001215\,mol$

$$= 0.01215\,mol \checkmark$$

The molar mass of the acid $H_2X = \dfrac{\text{mass of } H_2X \text{ used}}{\text{answer to (2)}} = \dfrac{1.53\,g}{0.01215\,mol}$

$$= 126\,g\,mol^{-1} \checkmark$$

Calculation of the number of molecules of water of crystallisation

This is the same as the method in the worked example ASC3, except that an extra step has to be carried out.

Worked example ASC4

4.11 g of hydrated sodium carbonate, $Na_2CO_3.xH_2O$, was weighed out and dissolved in water.

The solution was made up to $250\,cm^3$ and $25.0\,cm^3$ portions were titrated against standard $0.123\,mol\,dm^{-3}$ hydrochloric acid using methyl orange as the indicator. The mean titre was $23.45\,cm^3$.

The equation for the reaction is:

$$Na_2CO_3(aq) + 2HCl(aq) \rightarrow 2NaCl(aq) + CO_2(g) + 2H_2O(l)$$

Calculate:

(a) the amount of hydrochloric acid in the mean titre (1)

(b) the amount of sodium carbonate in $25.0\,cm^3$, and hence the amount in $250\,cm^3$ (2)

(c) the molar mass of the hydrated sodium carbonate using your answer to **(b)** and and the mass of hydrated sodium carbonate (1)

(d) the number of molecules of water of crystallisation in $Na_2CO_3.xH_2O$ using your answer to **(c)** (1)

[Relative atomic masses: H = 1; C = 12; O = 16; Na = 23]

Answer

(a) Amount of hydrochloric acid in the mean titre

$$= 0.123\,mol\,dm^{-3} \times \frac{23.45}{1000}\,dm^3$$

$$= 0.002884\,mol \checkmark$$

(b) Amount of sodium carbonate in $25.0\,cm^3 = \frac{1}{2} \times 0.002884\,mol$
$$= 0.001442\,mol \checkmark$$

Hence, the amount in $250\,cm^3 = 10 \times 0.001442\,mol$
$$= 0.01442\,mol \checkmark$$

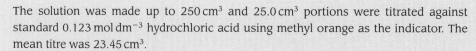 The moles of hydrochloric acid are multiplied by $\frac{1}{2}$ to get the moles of sodium carbonate because there is 1 Na_2CO_3 to 2 HCl in the equation.

(c) Molar mass of the hydrated sodium carbonate $= \dfrac{mass}{moles}$

$$= \frac{4.11\,g}{0.01442\,mol}$$

$$= 285\,g\,mol^{-1} \checkmark$$

(d) Number of molecules of water of crystallisation in $Na_2CO_3.xH_2O$.
mass of $\times\,H_2O = 285 - (mass\ of\ Na_2CO_3)$
$$= 285 - [(23 \times 2) + 12 + (3 \times 16)]$$
$$= 179$$

$$x = \frac{179\,g}{18\,g}$$
$$= 9.9$$

But the number of molecules must be a whole number, so the number of molecules of water of crystallisation = 10 $\checkmark$

Iodine/thiosulfate titrations

The method involves a solid oxidising agent being weighed out, dissolved in water and made up to $250 \, cm^3$. Alternatively an aqueous solution of an oxidising agent, such as hydrogen peroxide, H_2O_2, may be supplied.

$25 \, cm^3$ portions of the oxidising agent solution are pipetted into a conical flask and an approximate volume of dilute sulfuric acid added. Excess potassium iodide is then added — this reacts with the acidified oxidising agent to liberate iodine.

The iodine produced is then titrated against a standard solution of sodium thiosulfate until the colour fades to a pale yellow colour. Starch indicator is then added and the sodium thiosulfate added *drop by drop* until the blue–black starch/iodine complex goes colourless.

> If starch is added too early the complex will be formed irreversibly and the titre will be inaccurate.

The equation for the oxidation of iodide ions and that for the reaction between the liberated iodine and thiosulfate ions will be given.

The reaction between the oxidising agent, OA, and excess iodide ions is:

$$y \, OA(aq) + 2z \, I^-(aq) \rightarrow z \, I_2(aq) + \dots$$

So the amount of the oxidising agent $= \dfrac{y}{z} \times$ moles of I_2.

The reaction between the liberated iodine and thiosulfate ions is:

$$2S_2O_3^{2-}(aq) + I_2(aq) \rightarrow 2I^-(aq) + S_4O_6^{2-}(aq)$$

So the amount of $I_2 = \frac{1}{2} \times$ the moles $S_2O_3^{2-}$.

The calculation is best broken down into steps:

(1) Calculate the amount of sodium thiosulfate in the mean titre:

$$\text{moles of thiosulfate} = \text{concentration} \times \frac{\text{mean titre}}{1000}$$

(2) Calculate the amount of iodine liberated, and hence the number of moles of oxidising agent in $25.0 \, cm^3$:

$$\text{moles of iodine liberated} = \tfrac{1}{2} \times \text{moles of thiosulfate}$$

$$\text{moles of oxidising agent in } 25.0 \, cm^3 = \text{moles of iodine} \times \frac{y}{z}$$

Worked example ASC5

$0.85 \, g$ of a group 1 metal iodate(V), MIO_3, was weighed out, dissolved in water and made up to $250 \, cm^3$.

$25.0 \, cm^3$ portions were pipetted into a conical flask and excess potassium iodide and dilute sulfuric acid added.

The liberated iodine was titrated against $0.111 \, mol \, dm^{-3}$ sodium thiosulfate solution. The mean titre was $21.50 \, cm^3$.

The equations for the reactions are:

$$IO_3^-(aq) + 5I^-(aq) + 6H^+(aq) \rightarrow 3I_2(aq) + 3H_2O(l)$$

$$I_2(aq) + 2S_2O_3^{2-}(aq) \rightarrow 2I^-(aq) + S_4O_6^{2-}(aq)$$

Calculate:

(a) the number of moles of sodium thiosulfate in the mean titre (1)

(b) the number of moles of iodine liberated, and hence the number of moles of MIO_3 in $250 \, cm^3$, using your answer to **(a)** (3)

(c) the molar mass of MIO_3 using your answer to **(b)** and the mass of MIO_3 used (1)

(d) identify the element M in MIO_3 using your answer to **(c)** (1)

Relative atomic masses: $I = 126.9$; $O = 16.0$; $Li = 6.9$; $Na = 23.0$; $K = 39.1$; $Rb = 85.5$; $Cs = 132.9$

(e) How would the titre have been altered if the burette had been washed out with water and not with the sodium thiosulfate solution? (1)

Answer

(a) Number of moles of sodium thiosulfate in the mean titre = concentration × volume in dm^3

$$= 0.111 \, mol \, dm^{-3} \times \frac{21.50}{1000} \, dm^3$$

$$= 0.002387 \, mol \checkmark$$

(b) Number of moles of iodine liberated $= \frac{1}{2} \times 0.002387$
$$= 0.001193 \checkmark$$

Number of moles of MIO_3 in $25.0 \, cm^3 = \frac{1}{3} \times 0.001193$
$$= 0.0003978 \checkmark$$

Number of moles of MIO_3 in $250 \, cm^3 = 10 \times 0.0003978$
$$= 0.003978 \checkmark$$

(c) Molar mass of $MIO_3 = \dfrac{mass}{moles}$

$$= \frac{0.85 \, g}{0.003978 \, mol}$$

$$= 213.7 \, g \, mol^{-1} \checkmark$$

(d) A_r of $M = 213.7 - (126.9 + 3 \times 16.0)$
$$= 38.8$$

The group 1 metal M is potassium (relative atomic mass 39.1) $\checkmark$

(e) The solution in the burette would have been more dilute, so more would have had to be run in $\checkmark$

An alternative experiment provides you with a solution of an oxidising agent and you are asked to find its concentration.

Worked example ASC6

25.0 cm³ of a solution of hydrogen peroxide was pipetted into a conical flask. Excess potassium iodide and dilute sulfuric acid were added and the liberated iodine titrated against 0.111 mol dm⁻³ sodium thiosulfate solution.

The mean titre was 26.45 cm³.

The equations for the reactions are:

$$H_2O_2(aq) + 2H^+(aq) + 2I^-(aq) \rightarrow I_2(aq) + 2H_2O(aq)$$

$$I_2(aq) + 2S_2O_3^{2-}(aq) \rightarrow 2I^-(aq) + S_4O_6^{2-}(aq)$$

Calculate:

(a) the number of moles of sodium thiosulfate in the mean titre   (1)

(b) the number of moles of iodine, and hence the number of moles of hydrogen peroxide, in 25.0 cm³   (2)

(c) the concentration of the hydrogen peroxide solution in mol dm⁻³ using your answer to **(b)**   (1)

(d) the volume of oxygen gas liberated when 1.00 dm³ of this solution decomposes: (1)

$$2H_2O_2(aq) \rightarrow 2H_2O(l) + O_2(g)$$

1 mol of a gas occupies 24 dm³ under these conditions of temperature and pressure.

Answer

(a) Number of moles of sodium thiosulfate in the mean titre = concentration × volume in dm³

$$= 0.111 \text{ mol dm}^{-3} \times \frac{26.45}{1000} \text{ dm}^3$$

$$= 0.002936 \text{ mol} ✓$$

(b) Number of moles of iodine, $I_2 = \frac{1}{2} \times 0.002936$
$$= 0.001468 ✓$$

Hence the amount of hydrogen peroxide in 25.0 cm³ = 0.001468 mol ✓

(c) Concentration of the hydrogen peroxide solution $= \dfrac{\text{moles}}{\text{volume}}$

$$= \frac{0.001460}{0.0250 \text{ cm}^3}$$

$$= 0.05872 \text{ mol dm}^{-3} ✓$$

(d) Amount of oxygen produced $= \frac{1}{2} \times 0.05872$ mol
$$= 0.02936 \text{ mol}$$

Volume of oxygen produced $= 24 \text{ dm}^3 \text{ mol}^{-1} \times 0.02936 \text{ mol}$
$$= 0.705 \text{ dm}^3 ✓$$

Enthalpy change experiments

The keys to the calculations are:

- the heat change = mass of solution $\times$ 4.18 $\times$ ΔT

- the enthalpy change $\Delta H = \pm \dfrac{\text{heat change}}{\text{number of moles reacted}}$

- if the temperature rises, the reaction is exothermic and so ΔH is negative (and if it falls, ΔH is positive).

There are two types of experiment.

(1) Estimation of ΔH of a single reaction. A typical mark allocation is:

Recording volumes and temperatures	2 marks
Drawing and interpreting graph	3 or 4 marks
Accuracy	4 or 5 marks
Calculation	3 or 4 marks
Comment	1 mark

(2) Estimation of ΔH for two reactions and the use of Hess's law to calculate ΔH of a third reaction. The mark allocation for the two experiments is:

Recording masses and temperatures	2 marks each
ΔT accuracy	2 marks each
Calculation of ΔH	2 marks each
Final calculation using Hess's law	2 marks

Experimental methods

There are three different techniques — they all use a plastic (expanded polystyrene) cup held in a glass beaker with a thermometer being used to stir the reactants.

(1) Mix measured amounts (volumes or masses) and record the starting and maximum temperatures. This method is used for rapid reactions, such as acid/base or acid/carbonate.

(2) Record the temperature of one solution for several minutes, then add the other reagent and measure the temperature over a period of time. A graph of temperature against time is then plotted. This method is used for slow reactions, such as metal/copper sulfate solution, or to measure the enthalpy of solution of salts.

(3) Measure the initial temperature of one solution and add portions of the other solution at regular intervals, measuring the temperature each time. A graph of temperature against volume added is then plotted. This method is used for rapid reactions, such as acid/base reactions.

Errors in technique

- Not using the bottom of the meniscus to read volumes.
- Failing to stir the mixture thoroughly, and so not observing the correct temperature.

- Failing to read the temperature to sufficient accuracy. You are advised to read all temperatures to an accuracy of $\pm0.1°C$, so you must be prepared to estimate it between graduations if you are supplied with a thermometer calibrated to 0.2°C. The temperature in the diagram is 34.3°C.

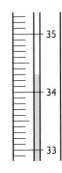

- Careless addition of solid. If a gas is evolved, as in the reaction between a carbonate and an acid, care must be taken to add the solid in small portions to prevent any solution frothing up out of the plastic cup.

Errors in recording
- Failing to record all masses to 0.01 g. For example, a mass of 12.30 g must not be recorded as 12.3 g.
- Failing to record temperature to 1 decimal place. You must record all temperatures to 0.1°C.
- You must record times to an accuracy of at least 1 minute.

Errors in graph plotting
- A sensible scale must be chosen. Make sure that at least half of each axis is used for the readings. Do **not** start the temperature axis at zero degrees.
- Check that you have plotted each point accurately.
- Do not forget to label both axes.
- Make sure that two best-fit straight lines are drawn using a ruler. Do **not** join up the points.
- Extrapolate both lines backwards, so as to be able to measure a maximum ΔT or the volume required for neutralisation — see worked examples ASC7 and ASC8.

Calculation
- The first step is to calculate the heat change:

heat change (in joules) = mass of solution $\times$ specific heat capacity $\times \Delta T$

$$\text{heat change in kJ} = \frac{\text{heat change in J}}{1000}$$

The specific heat capacity will always be given. A common error is to use the mass of a reagent and not the mass of solution in your calculation — see worked example ASC7.

Assume that the density of all aqueous solutions is $1\,g\,cm^{-3}$. This means that the mass is numerically equal to the volume.

- The second step is to calculate the number of moles reacted. For a solid this will be:

$$\frac{\text{mass}}{\text{molar mass}}$$

For a solution it will be:

concentration $\times$ volume in dm^3

- The third step is to calculate ΔH:

$$\Delta H = \pm \frac{\text{heat change in kJ}}{\text{number of moles reacting}}$$

ΔH will be negative if the reaction is exothermic (the temperature rises) and positive if the reaction is endothermic (the temperature falls).

It is a common error to ignore the sign of ΔH or to get it wrong.

If you are asked to give your answer to ΔH to 2 significant figures, make sure that you do so and do not give it to 3 or more (see below).

Comment on accuracy

You may be asked to comment about which is the least accurate piece of equipment — this will usually be the thermometer. For a thermometer reading to 0.1°C for each reading, the error for each reading is ±0.1°C. As two readings were taken to measure ΔT, the percentage error is $\frac{2 \times 0.1}{\Delta T} \times 100\%$.

The appropriate number of significant figures is that of the number of significant figures in the value of ΔT. If this is less than 10°, for example a temperature change of 7.6°C, give your answer to 2 s.f. If it is over 10°, for example a change of 23.4°C, you should give your answer to 3 s.f.

Increasing the volume of solution that is used will increase the heat produced, but will proportionately increase the mass of solution that is heated up. So it will have no effect on ΔT and hence on the accuracy. Increasing the concentration of a solution or the mass of a solid will increase the value of ΔT and hence increase the accuracy.

The following is an example of a rapid reaction using the technique that involves measuring the initial and final temperatures.

Worked example ASC7

When 2.58 g of potassium carbonate, K_2CO_3, was carefully added to 25.0 cm^3 (an excess) of 2.0 mol dm^{-3} hydrochloric acid, the temperature rose from 23.7°C to 31.4°C:

$$K_2CO_3(s) + 2HCl(aq) \rightarrow 2KCl(aq) + CO_2(g) + H_2O(l)$$

Assuming that the specific heat capacity of the solution is 4.18 J g^{-1} °C^{-1}, calculate:

(a) the heat energy transferred (1)

(b) the enthalpy change, ΔH_1, giving your answer to an appropriate number of significant figures (2)

Answer

(a) Energy transferred $=$ mass of solution $\times 4.18\,\text{J}\,\text{g}^{-1}\,°\text{C}^{-1} \times \Delta T$

$$= 25.0\,\text{g} \times 4.18\,\text{J}\,\text{g}^{-1}\,°\text{C}^{-1} \times 7.7°\text{C}$$

$$= 805\,\text{J}$$

$$= 0.805\,\text{kJ} \checkmark$$

e Make sure that you use the mass of the solution (25 g) and not the mass of potassium carbonate solid taken (2.58 g) in this calculation.

(b) Amount of $K_2CO_3 = \dfrac{\text{mass}}{\text{molar mass}}$

$$= \dfrac{2.58\,\text{g}}{138.2\,\text{g}\,\text{mol}^{-1}}$$

$$= 0.0187\,\text{mol} \checkmark$$

$$\Delta H_1 = \dfrac{-0.805\,\text{kJ}}{0.0187\,\text{mol}}$$

$$= -43\,\text{kJ}\,\text{mol}^{-1} \checkmark$$

e The value of ΔT is only to 2 significant figures and so the answer to ΔH must also be to 2 s.f. The sign of ΔH is negative because the temperature rose.

The following is an example of a Hess's law calculation.

Worked example ASC8

Potassium hydrogencarbonate also reacts with acids:

$$KHCO_3(s) + HCl(aq) \rightarrow KCl(aq) + CO_2(g) + H_2O(l)$$

Given that the enthalpy change, ΔH_2, for this reaction is $+29\,\text{kJ}\,\text{mol}^{-1}$ and using your answer to **ASC7 (b)**, draw a Hess's law diagram and calculate the value of ΔH for the reaction:

$$2KHCO_3(s) \rightarrow K_2CO_3(s) + CO_2(g) + H_2O(l)$$

Answer

Labelled diagram $\checkmark\checkmark$

$$\Delta H = (2 \times \Delta H_2) - \Delta H_1$$

$$= 2 \times (+29\,\text{kJ}\,\text{mol}^{-1}) - (-43\,\text{kJ}\,\text{mol}^{-1})$$

$$= +101\,\text{kJ}\,\text{mol}^{-1} \checkmark$$

The example below is for a slow reaction.

Worked example ASC9

25.0 cm^3 of 0.500 mol dm^{-3} copper(II) sulfate solution was poured into a polystyrene cup and its temperature measured every minute for 3 minutes.

At the fourth minute, 1.3 g (an excess) of zinc was added and the temperature measured for a further 6 minutes:

$$Zn(s) + CuSO_4(aq) \rightarrow ZnSO_4(aq) + Cu(s)$$

The results are shown below.

Time/min	0	1	2	3	4	5	6	7	8	9	10
Temperature/°C	22.0	22.0	22.0	22.0		29.3	30.7	30.2	29.8	29.4	29.0

(a) Draw a graph of temperature on the *y*-axis against time on the *x*-axis. (1)

(b) Draw a line of best fit from zero to the third minute, and another from the sixth to the tenth minute. Extrapolate the lines to the fourth minute and calculate the estimated temperature rise. (max. 5)

(c) Calculate the heat change, assuming that the solution has a specific heat capacity of 4.18 J K^{-1}°C^{-1} and a density of 1 g cm^{-3}. (1)

(d) Hence calculate the enthalpy change to 2 significant figures for the reaction, giving a sign and units with your answer. (3)

(e) What observation shows that the zinc was in excess? (1)

(f) Would doubling all quantities increase the accuracy of your result? Justify your answer. (2)

Answer

(a) Labelled axes with sensible scale and points plotted correctly ✓

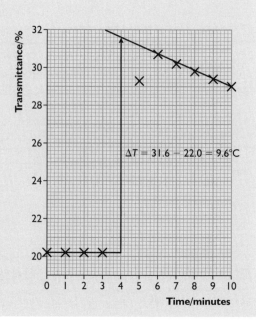

$\Delta T = 31.6 - 22.0 = 9.6°C$

(b) Two straight lines drawn and extrapolated ✓; from graph: $\Delta T = 9.6°C$ ✓; accuracy value $\pm0.5°C$ ✓✓✓; $\pm1.0°C$ ✓✓; $\pm2.0°C$ ✓

(c) Heat change = mass of solution × specific heat capacity × ΔT

$$= 25.0\,g \times 4.18\,J\,g^{-1}°C^{-1} \times 9.6°C$$
$$= 1003\,J$$
$$= 1.003\,kJ ✓$$

(d) Amount of copper(II) sulfate = $0.500\,mol\,dm^{-3} \times 0.025\,dm^{-3}$
$$= 0.0125\,mol ✓$$

$$\Delta H = \frac{-1.003\,kJ}{0.0125\,mol}$$

$$= -80.24\,kJ\,mol^{-1} \text{ (the temperature rose)}$$
$$= -80\,kJ\,mol^{-1} ✓ \text{ for value to 2 s.f.;} ✓ \text{ for sign and mark}$$

(e) The solution would become colourless ✓

(f) No, because the amount of heat would double, but so would the mass of solution to be heated up ✓
Thus ΔT would not alter and hence the accuracy would also not alter ✓

The next example is for steady addition of a reactant in a rapid reaction.

Worked example ASC10

Solutions of nitric acid and sodium hydroxide were allowed to reach the same temperature of 18.0°C.

50.0 cm³ of the nitric acid was pipetted into a polystyrene cup and a 2.0 mol dm⁻³ solution of sodium hydroxide was added, with stirring, in 5.0 cm³ portions. The temperature was noted after each addition.

The reaction is:

$$HNO_3(aq) + NaOH(aq) \rightarrow NaNO_3(aq) + H_2O(l)$$

The results are shown below.

Volume of NaOH/cm³	0	5	10	15	20	25	30	35	40	45
Temperature/°C	18.0	19.8	21.5	23.5	25.0	26.8	27.2	27.0	26.7	26.5

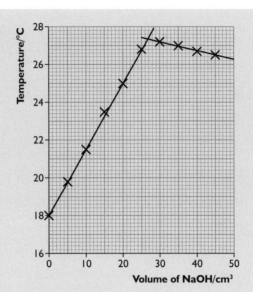

Volume of NaOH/cm³

(a) Use the graph to calculate ΔT for the reaction and the volume of sodium hydroxide that exactly neutralises the nitric acid. (7)

(b) Use your answer to (a) to calculate the volume in the polystyrene cup when the nitric acid had just been neutralised. (1)

(c) Calculate the heat change for the neutralisation of the nitric acid. (1)

(d) Use the volume calculated in (a) to calculate the number of moles of sodium hydroxide required to neutralise the nitric acid, and hence the number of moles of the acid. (1)

(e) Use your answers to (c) and (d) to calculate the enthalpy of neutralisation of nitric acid. (1)

(f) Is this as accurate a method for finding the concentration of an acid compared with a standard acid/base titration? (1)

Answer

(a) From the graph: $\Delta T = 27.4 - 18.0$
$$= 9.4°C ✓$$
Volume of NaOH $= 26.5\,cm^3$ ✓
Accuracy of ΔT: $\pm 0.5°C$ ✓✓✓; $\pm 1.0°C$ ✓✓; $\pm 2.0°C$ ✓
Accuracy of volume: $\pm 0.5\,cm^3$ ✓✓; $\pm 1.0\,cm^3$ ✓

(b) Total volume $= 50\,cm^3 + 26.5\,cm^3$
$$= 76.5\,cm^3 ✓$$

(c) Heat change $= 76.5\,g \times 4.18\,J\,g^{-1}°C^{-1} \times 9.4\,°C^{-1}$
$$= 3006\,J$$
$$= 3.006\,kJ ✓$$

(d) Amount of NaOH = 2.0 mol dm^{-3} × 0.0265 dm^{-3}

= 0.053 mol

= number of moles of HNO$_3$ ✓

(e) $\Delta H = \dfrac{-3.006\,\text{kJ}}{0.053\,\text{mol}}$

= $-57\,\text{kJ}\,\text{mol}^{-1}$ ✓

(f) No, because the volume cannot be estimated to better than ±0.5 cm^3 ✓

Activity D: Preparation

In all preparations you will be asked to work out the percentage yield. This will score 2 or 3 of the 12 marks in the assessment.

The method is:

(1) Work out the molar mass of the reactant that you measured out and of the product.

(2) Use the expression moles = $\dfrac{\text{mass}}{\text{molar mass}}$ to calculate the number of moles of reactant.

(3) If the reactant and product are in a 1 : 1 ratio in the equation, the **theoretical** moles of product = moles of reactant.

(4) Theoretical mass of product = moles × molar mass.

(5) % yield = $\dfrac{\text{actual mass of product}}{\text{theoretical mass of product}}$ × 100%

The percentage yield is **not** $\dfrac{\text{mass of product}}{\text{mass of reactant}}$ × 100%.

Inorganic preparations

You will be required to prepare either a simple salt such as copper(II) sulfate or nickel(II) sulfate, or a double salt such as ammonium iron(II) sulfate. You will be given detailed instructions. Any salt you prepare will contain water of crystallisation.

Preparation of a simple salt

The solid reactant is added a little at a time to a known volume of acid, which has been gently heated.

When there is a slight excess of solid, the solution is filtered into an evaporating basin and the filtrate evaporated until about half of its initial volume remains.

The solution is cooled overnight and the crystals dried by pressing between sheets of filter paper. The crystals are then weighed.

A typical mark scheme is:

Recording the masses	1 or 2 marks
Observations and comments on technique	3 or 4 marks
Description of the appearance of product	2 marks
Calculation of percentage yield	2 marks
Value of % yield	2 marks if >50% 1 mark if between 49 and 25%
Comment as to why yield is <100%	1 mark

Errors in technique
- Not having the acid solution hot enough.
- Adding the solid too fast so that it froths up and overflows.
- Evaporating the solution too far so that some anhydrous salt is formed, or not far enough so that too much product remains in solution.

Errors in recording
- All the masses must be given to 2 decimal places.
- The description of the crystals must include their colour and their shape, for instance green and needle or diamond shaped.

Worked example ASD1

Using a measuring cylinder, $25 \, cm^3$ of $1.0 \, mol \, dm^{-3}$ sulfuric acid was transferred to a beaker and the solution warmed. Portions of copper(II) oxide were added with stirring until the base was in excess.

The hot solution was filtered and the solution evaporated to half its volume. The solution was left to cool. The crystals formed were dried and weighed in a pre-weighed weighing bottle.

Mass of container + hydrated copper(II) sulfate	29.66 g
Mass of empty container	24.65 g
Mass of hydrated copper(II) sulfate	5.01 g

(a) What was observed as the copper(II) oxide was added? (2)

(b) What showed that the copper oxide was in excess? (1)

(c) Why must the solution not be evaporated to dryness? (1)

(d) Describe the appearance of the crystals. (2)

(e) Calculate the mass of $CuSO_4.5H_2O$ ($M_r = 249.6$) that could be made from $25 \, cm^3$ of $1.0 \, mol \, dm^{-3}$ sulfuric acid. (2)

(f) Calculate the percentage yield of the preparation. (max. 3)

(g) Why is the yield less than 100%? (2)

Answers

(a) Slight frothing ✓ and the solution turning blue ✓

(b) Some of it remained undissolved ✓

(c) If evaporated to dryness, the anhydrous salt will be formed ✓

(d) Blue ✓ and rhombus shaped ✓

(e) Number of moles of acid $= 1.0\,\text{mol dm}^{-3} \times 0.025\,\text{dm}^3$
$$= 0.025\,\text{mol}$$
Theoretical amount of $CuSO_4.5H_2O = 0.025\,\text{mol}$ ✓
Theoretical mass of $CuSO_4.5H_2O = \text{moles} \times \text{molar mass}$
$$= 0.025\,\text{mol} \times 249.6\,\text{g mol}^{-1}$$
$$= 6.24\,\text{g}\ \checkmark$$

(f) $\%\ \text{yield} = \dfrac{\text{actual mass of product}}{\text{theoretical mass of product}} \times 100\%$

$$= \dfrac{5.01\,\text{g}}{6.24\,\text{g}} \times 100\%$$

$$= 80\%\ \checkmark\ \text{plus}\ \checkmark\checkmark\ \text{for} >50\%\ \text{obtained}$$

(g) Some of the saturated solution did not crystallise ✓; or if no solution was left after cooling the answer would be that some anhydrous salt was formed ✓

Preparation of a double salt

The most likely double salt to be used is hydrated ammonium iron(II) sulfate.

The preparation is in three steps, and the details for each will be given clearly.

(1) *Preparation of a solution of iron(II) sulfate.* A known mass of iron filings is reacted with a slight excess of hot dilute sulfuric acid. Any unreacted iron is filtered off.

(2) *Preparation of an ammonium sulfate solution.* Dilute ammonia is added in portions to dilute sulfuric acid until the solution is alkaline to litmus. The excess ammonia is then boiled off.

(3) *Mixing the solutions and carrying out partial evaporation.* The two solutions are mixed in a beaker and the mixture carefully evaporated until the volume has halved. It is then left to cool and the crystals are filtered, dried and weighed.

A typical mark scheme is:

Weighing the iron and the product	2 marks
Observations and comments on technique	3 or 4 marks
Appearance of crystals	2 marks
Calculation of percentage yield	2 marks
Value of % yield	2 marks if >50% 1 if between 49 and 25%
Comment as to why yield is <100%	1 mark

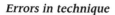

Errors in technique

- Not having the acid solution hot enough for the reaction with iron.
- Adding the solid too quickly so that it froths up and overflows.
- Not stirring the acid solution when adding ammonia.
- Evaporating the solution too far, so that some anhydrous salt is formed.

Errors in recording

- All the masses must be given to 2 decimal places.
- The description of the crystals must include their colour and their shape, for instance green and needle shaped or diamond shaped.

Worked example ASD2

25 cm³ of dilute sulfuric acid was placed in a beaker and heated to just below boiling. 2.75 g of iron was added in small portions, and after each addition a plug of cotton wool was placed in the neck of the flask. Finally an extra 3 cm³ of acid was added and the solution filtered.

25 cm³ of dilute sulfuric acid was added to a beaker and 30 cm³ of ammonia solution added and stirred. Further 5 cm³ portions of ammonia were then added until a drop of the solution (removed with a glass rod) turned red litmus blue. The solution was boiled to remove excess ammonia.

The two solutions were mixed and then boiled until the volume had been reduced by half. It was left to cool and the crystals that formed were filtered off, dried and weighed. They weighed 15.2 g.

(a) Why must the iron not be added all at once? (1)

(b) Why was a plug of cotton wool placed in the neck of the flask? (1)

(c) In the preparation of the second solution, was the ammonia solution in excess? Justify your answer. (1)

(d) Describe the appearance of the crystals (2)

(e) Calculate the theoretical yield of ammonium iron(II) sulfate, $(NH_4)_2SO_4.FeSO_4.6H_2O$ (molar mass = 391.8 g mol⁻¹), assuming that 1 mol of iron produces 1 mol of the hydrated double salt. (2)

(f) Calculate the percentage yield of product. (1)

(g) Why is the yield less than 100%? (2)

Answers

(a) Because the acid would froth up and overflow ✓

(b) To prevent acid spray from escaping ✓

(c) Yes, because the litmus turned blue ✓

(d) Pale green ✓ and diamond shaped ✓

(e) Amount of iron $= \dfrac{2.75\,g}{55.8\,g\,mol^{-1}}$

$$= 0.0493\,mol\ ✓$$

Theoretical mass of product $= 0.0493\,mol \times 391.8\,g\,mol^{-1}$
$$= 19.31\,g\ ✓$$

(f) % yield $= \dfrac{15.2\,g}{19.31\,g} \times 100\%$

$$= 79\%\ ✓$$

(g) Some of the saturated solution did not crystallise ✓; or if no solution was left after cooling the answer would be that some anhydrous salt was formed ✓

Organic preparations

Full details of the procedure will be given and they should be followed carefully.

Technique

- You must know how to set up apparatus for distillation while adding reagent, and for distillation.
- Make sure that the water flows into the condenser at the bottom and out at the top.
- Check that all the joints are sound and that they will not leak.
- Check that a thermometer, if fitted, is opposite the entrance of the condenser.

Oxidation of a primary or secondary alcohol

A known volume of the alcohol is dissolved in a solution of potassium dichromate(VI). Very carefully, some concentrated sulfuric acid is added to some water in a flask. This is then connected to distillation apparatus that is adapted for the addition of a reagent and brought to the boil.

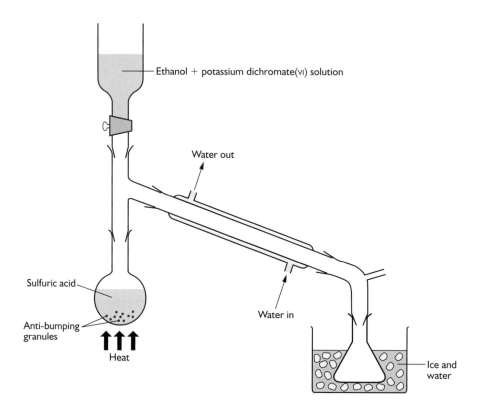

The alcohol and potassium dichromate(VI) solution is then added from the tap funnel a few drops at a time.

A primary alcohol is oxidised to an aldehyde, which distils over as it is formed and is collected in a flask surrounded by iced water:

$$CH_3CH_2OH + [O] \rightarrow CH_3CHO + H_2O$$

If heated under reflux, a primary alcohol would be oxidised to a carboxylic acid:

$$CH_3CH_2OH + 2[O] \rightarrow CH_3COOH + H_2O$$

A secondary alcohol is oxidised to a ketone. This also is distilled off and collected in a flask surrounded by iced water:

$$CH_3CH(OH)CH_3 + [O] \rightarrow CH_3COCH_3 + H_2O$$

Aldehydes and ketones can be tested for by adding a solution of 2,4-dinitrophenyl-hydrazine. They give a yellow or orange precipitate.

Aldehydes can be tested for by adding to:
- Tollens' reagent, which gives a silver mirror on warming
- Fehling's (or Benedict's) solution, which gives a red precipitate on warming.

Carboxylic acids react with sodium carbonate (or hydrogencarbonate) to evolve a gas which turns limewater cloudy.

A typical mark scheme is:

Observation of test on product	2 marks
Observations during oxidation	1 mark
Description of product	1 or 2 marks
Boiling temperature of product	2 marks*
Discussion of technique	3 or 4 marks
Modification of apparatus	2 or 3 marks†

*Only if the product is redistilled
†Only when a primary alcohol is used

Worked example ASD3

Ethanol was oxidised as detailed above. The product was tested by adding some 2,4-dinitrophenylhydrazine solution.

(a) What was observed on addition of 2,4-dinitrophenylhydrazine solution? (2)

(b) Why was the sulfuric acid added to the water, and not the water to the acid? (1)

(c) What colour change took place during the oxidation? (2)

(d) Describe the product collected. (1)

(e) Why is it necessary to have a vent on the adapter connected to the collecting vessel? (1)

(f) Why must the ethanol/dichromate mixture be added a few drops at a time? (2)

(g) How would the apparatus be adapted to prepare ethanoic acid from ethanol? (2)

Answers
(a) A yellow ✓ (or orange) precipitate ✓

(b) Because the heat produced would vaporise the water and acid steam would be produced ✓

(c) The orange solution ✓ turned green ✓

(d) It is a colourless liquid ✓

(e) To prevent a build-up of pressure ✓

(f) The reaction is exothermic and so would become too vigorous ✓ and would froth up into the condenser ✓

(g) The apparatus would be arranged for heating under reflux ✓
The ethanoic acid would then be distilled out of the mixture ✓

Dehydration of an alcohol

An alcohol, such as cyclohexanol, can be dehydrated using concentrated phosphoric acid, producing an alkene.

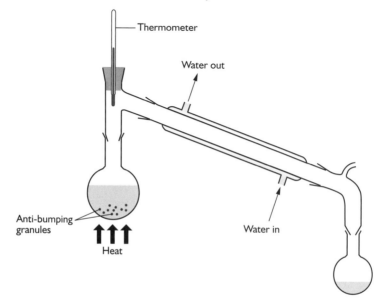

The acid and the alcohol are mixed and the cycloalkene is distilled off.

The product can then be tested with bromine water.

Worked example ASD4

Some concentrated phosphoric acid, a measured volume of cyclohexanol and a few anti-bumping granules are placed in a flask which is then attached to distillation apparatus.

The flask is warmed and a mixture of water and cyclohexene distils off and is collected. The mixture is poured into a separating funnel and the lower aqueous layer discarded. The upper layer is run into a flask and lumps of calcium chloride added. After a few minutes, the cyclohexene is poured off and distilled.

(a) What is observed when some of the cyclohexene is shaken with bromine water? Write the equation for the reaction. (3)

(b) Why is the cyclohexene in the upper layer in the separating funnel? (1)

(c) What is the function of the calcium chloride? (1)

(d) Why must the thermometer bulb be placed opposite the mouth of the condenser? (1)

Answers

(a) The bromine water turns from a red-brown colour to colourless ✓ and two layers form ✓

(cyclohexene) + Br_2 + H_2O ⟶ (2-bromocyclohexanol, with OH and Br) + HBr ✓

🖉 The equation below would also be allowed.

(cyclohexene) + Br_2 ⟶ (1,2-dibromocyclohexane, with Br and Br)

(b) Because it is less dense than water ✓

(c) To dry the cyclohexene ✓

(d) Above or below this point, the temperature would not be the boiling temperature of the distillate ✓

Unit 6

Activity B: Qualitative observation

Organic substances

Introduction

It is important to read the student brief carefully. Hints may be given such as:

- each organic compound contains only one functional group
- the number of carbon atoms that molecules of each compound contains and whether it is a branched or a straight-chain compound
- the fact that one unknown may be oxidised, reduced or hydrolysed to another of the unknown or unknowns

You will always be given the quantities of both the unknown and the test reagent and full details of how to carry out the test, so you do not need to learn these.

Although the main part of the task will be about substances met in Unit 4, questions will be asked about alkenes, alcohols or possibly halogenoalkanes, so these must also be revised before the practical assessment.

Make sure that you can interpret mass, infrared and NMR spectra before you do the assessment.

Make sure that you know the difference between **displayed** formulae (all the atoms and bonds must be shown), **structural** formulae (an unambiguous formula showing each group separately) and **skeletal** formulae.

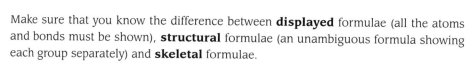

| Displayed | Structural | Skeletal |

Make sure that your deductions are logical. The deduction from a positive iodoform test is not that the unknown is one of ethanal, a methyl ketone or an alcohol with a $CH_3CH(OH)$ group if you have already shown that an unknown is a ketone.

The unknowns that you could be asked to identify are:
- alkenes
- alcohols
- halogenoalkanes (not very likely)
- aldehydes and ketones
- carboxylic acids
- carboxylic acid derivatives (not very likely).

Possible observations marks are shown with a ✓.

A typical mark scheme is:

Observations	8 or 9 marks
Analysis of spectra	2 or 3 marks
Identification of unknowns	2 or 3 marks
Comment on product of reaction	1 mark

Possible tests

Combustion

Test	Observations	Inferences
Burn a few drops on a crucible lid	• It burns with a clear non-smoky flame ✓ • It burns with a smoky flame ✓	• Low carbon to hydrogen ratio • High carbon to hydrogen ratio

Solubility in water and pH test

Test	Observations	Inferences
Add to water	• Either two layers form ✓;	• Either no OH group; or if an alcohol or acid it has a high molar mass (at least four carbon atoms)
	• Or it dissolves fully ✓	• Forms hydrogen bonds with water, so has an OH group
Add litmus	• Litmus paper (or UI solution) goes red ✓	• It is an acid
or universal indicator (UI)	• UI solution goes green ✓	• It is a neutral substance (not an acid)

Test for alkenes

Test	Observations	Inferences
Add some bromine water to the unknown — stopper the test tube and shake carefully	Brown bromine water goes colourless ✓ and two layers are formed ✓	C=C group present

Tests for halogenoalkanes

Test	Observations	Inferences
Shake a few drops of the unknown with water	Two layers form ✓	
Add a few drops of ethanol and aqueous silver nitrate and dilute nitric acid — stand the test tube in a beaker of hot water	• White precipitate ✓ • Cream precipitate ✓ • Yellow precipitate ✓	• It contains a C–Cl group • It contains a C–Br group • It contains a C–I group
Add 4 drops of ethanol and then 2 cm³ of dilute sodium hydroxide — stand the test tube in a beaker of hot water for 5 minutes; then add excess nitric acid followed by aqueous silver nitrate	• White precipitate ✓ • Cream precipitate ✓ • Yellow precipitate ✓	• It contains a C–Cl group • It contains a C–Br group • It contains a C–I group
Add concentrated sodium hydroxide in ethanol — fit the test tube with a delivery tube and warm gently, passing any gas evolved through a little bromine water	• Brown bromine goes colourless ✓	• H–halogen eliminated forming an alkene

The identity of the halogen in the precipitate formed with silver nitrate can be confirmed by adding ammonia solution.

Observation	Inference
Precipitate dissolves in dilute ammonia ✓	The unknown is a chloroalkane
Precipitate insoluble in dilute but soluble in concentrated ammonia ✓	The unknown is a bromoalkane
Precipitate insoluble in concentrated ammonia ✓	The unknown is an iodoalkane

Test for aldehydes and ketones
These contain the C=O group.

Test	Observations	Inferences
Add a few drops of a solution of 2,4-dinitrophenylhydrazine	A yellow or orange precipitate ✓	It is an aldehyde or a ketone

Tests to distinguish between an aldehyde and a ketone

Test	Observations	Inferences
Add a few drops of the unknown to some Fehling's (or Benedict's) solution and warm	• Either red precipitate ✓ • Or the blue colour remains ✓	• It is an aldehyde • It is a ketone*
Add a few drops of the unknown to Tollens' reagent (see below) and warm	• Either silver mirror formed ✓ • Or solution stays colourless ✓	• It is an aldehyde • It is a ketone*
Add a few drops of the unknown to acidified potassium dichromate(VI) solution	• Either orange solution goes green ✓ • Or solution stays orange ✓	• It is an aldehyde* • It is a ketone*
Add a few drops of the unknown to acidified potassium manganate(VII) solution and warm	• Either purple solution decolourised ✓ • Or solution stays purple ✓	• It is an aldehyde* • It is a ketone*

*This inference can only be made if the substance has already been shown to be a carbonyl compound.

Tollens' reagent is made by adding aqueous sodium hydroxide drop by drop to aqueous silver nitrate until a grey precipitate is formed. The clear liquid above the precipitate (the supernatant liquid) is poured off and the precipitate dissolved in the minimum of dilute ammonia.

The iodoform test

This is a reaction between an organic alcohol or carbonyl compound and a solution of iodine and sodium hydroxide. The mixture is then allowed to stand.

Observation	Previous test	Inferences
Pale yellow precipitate (of iodoform*) ✓	• A precipitate with 2,4-dinitrophenylhydrazine — showing it to be a carbonyl compound	• It contains a $CH_3C=O$ group
	• A red precipitate with Fehling's (or Benedict's) solution or a silver mirror with Tollens' reagent — showing it to be an aldehyde	• It is ethanal
	• Steamy fumes with PCl_5 or other test — showing it to be an alcohol (see below)	• It contains the $CH_3CH(OH)$ group, for example propan-2-ol

*Iodoform's systematic name is triiodomethane, CHI_3.

Tests for OH group in alcohols and carboxylic acids

Test	Observations	Inferences
Add a small piece of sodium to the unknown in an evaporating basin	• Bubbles evolved ✓ • Sodium disappears or a white solid forms ✓	• It contains an OH group and so is an alcohol or a carboxylic acid
Add solid phosphorus(V) chloride and test any gas evolved: either with a glass rod dipped in concentrated ammonia; or with damp blue litmus paper	• Steamy fumes evolved ✓ • White smoke formed ✓ • Litmus goes red ✓	• It contains an OH group and so is an alcohol or a carboxylic acid

To distinguish an acid from an alcohol

Test	Observations	Inferences
Add the unknown to a solution of sodium carbonate or sodium hydrogencarbonate and test any gas evolved with limewater	• Fizzing ✓ • Gas evolved turns limewater cloudy ✓	• It is a carboxylic acid
Add blue litmus (or universal indicator) to a solution of the unknown	• Litmus goes red ✓	• It is a carboxylic acid
Add some ethanol (or other alcohol) and a few drops of concentrated sulfuric acid to the unknown and warm. Then pour into a beaker containing some sodium carbonate solution and cautiously smell the product	• Fizzing ✓ • Smells of glue/fruity smell ✓	• It is a carboxylic acid (and an ester is formed)

To distinguish an alcohol from a carboxylic acid

	Observations	Inferences
Add some ethanoic acid and a few drops of concentrated sulfuric acid to the unknown and warm. Pour into a beaker containing some sodium carbonate solution and cautiously smell the product	• Fizzing ✓ • Smells of glue/fruity smell ✓	• It is an alcohol (and an ester is formed)
Warm the unknown gently with acidified potassium dichromate(VI) solution	• Orange solution goes green ✓	• It is a primary or secondary alcohol*

*This inference can only be made if it has been shown that the unknown contains an OH group. This is because aldehydes also turn acidified dichromate(VI) green.

A substance that has an OH group, forms an ester with ethanoic acid but does not change the colour of acidified potassium dichromate(VI), must be a tertiary alcohol.

Tests for acid chlorides

Test	Observations
Add the unknown to some water in a test tube Then add dilute nitric acid and aqueous silver nitrate	• Steamy fumes ✓ • White precipitate ✓
Add the unknown to some ethanol and then pour into a beaker containing aqueous sodium carbonate. Carefully smell the beaker's contents	• Fruity/glue smell ✓

Spectra

Mass spectra

You may be given a mass spectrum to help identify the unknown. The m/e values to look for are:

- the largest value is due to the molecular ion, M^+, and corresponds to the relative molecular mass of the compound
- a peak at $(M - 15)$ or at 15 due to the presence of a CH_3 group in the molecule
- a peak at $(M - 29)$ or at 29 due to either a C_2H_5 or a CHO group in the molecule
- a peak at $(M - 43)$ or at 43 due to a C_3H_7 group in the molecule
- a peak at $(M - 45)$ or at 45 due to a COOH group in the molecule

Infrared spectra

You will be allowed to use the *Edexcel data booklet* during the assessment, but note that it states that the C=O absorption in ketones is in the range $1680–700\,cm^{-1}$. This is wrong because only aromatic ketones fall in this range. Aliphatic ketones are in the range $1710–1720\,cm^{-1}$ and aliphatic aldehydes in the range $1720–1730\,cm^{-1}$.

Look out for absorptions in the range:

- $1710–1730\,cm^{-1}$ caused by the C=O group in aldehydes, ketones and carboxylic acids
- $1735–1750\,cm^{-1}$ caused by the C=O group in esters
- $3000–3400\,cm^{-1}$ caused by (hydrogen-bonded) O–H in alcohols
- $2800–3200\,cm^{-1}$ due to (hydrogen-bonded) O–H in carboxylic acids.

The hydrogen-bonded O–H absorption is very broad and it is difficult to assign a definite wavenumber.

NMR spectra

You will be allowed to use the *Edexcel data booklet* during the assessment.

The points to look for are:

- *The number of peaks* — this gives the number of different environments of the hydrogen atoms. For example, propan-2-ol, $CH_3CH(OH)CH_3$, will have three peaks because the hydrogen atoms in both CH_3 groups are in the same environment. Propan-1-ol, $CH_3CH_2CH_2OH$, will have four peaks.
- *The splitting pattern* — if a peak is not split, either the group has no hydrogen atoms on adjacent carbon atoms (as in propanone) or it is next to an O–H group:

– if it is split into 2 there is one hydrogen atom on the adjacent carbon atom
– if it is split into 3 then there are two hydrogen atoms on adjacent carbon atoms
– if it is split into 4 then there are three hydrogen atoms on adjacent carbon atoms

• *The chemical shift, δ ppm* — the H of the O–H in alcohols has a δ between 2 and 4 ppm:
  – in carboxylic acids between 11 and 12 ppm
  – in CH_3, CH_2 or CH between 0.5 and 3.5 ppm
  See the *Edexcel data booklet* for δ values.

Practice example A2B1

Compounds A and B both have molecules containing 3 carbon atoms and 1 oxygen atom.

Assume that A is propanone and that B is propan-2-ol. Write down the observations that would be made in the following tests. The answers are on pp. 85–86.

(a) Some 2,4-dinitrophenylhydrazine was added to each. Write down your observations with A and with B. (2)

(b) Some phosphorus(V) chloride was added to each. Write down your observations with A and with B. (2)

(c) Each was warmed with acidified potassium dichromate(VI) solution. Write down your observations with A and with B. (2)

(d) How many peaks would there be in the NMR spectrum of compound A, and how would they be split? (2)

Practice example A2B2

The molecules of compounds P and Q contain 4 carbon atoms, and molecules of compound R contain 3 carbon atoms.

Assume that P is butan-2-ol, Q is butanone and that R is propanoic acid. Write down the observations and deductions that would be made in the following tests. The answers are on p. 86.

(a) A small piece of sodium was added to P in an evaporating basin. Write down your observations. (2)

(b) P was warmed with acidified potassium dichromate(VI) solution. Write down your observations. (1)

(c) P was warmed gently with a mixture of iodine and sodium hydroxide solution. Write down your observations. (1)

(d) Use your answers to **(a)**, **(b)**, **(c)** and the information in the stem of the question to identify compound P by writing its name or formula. (1)

(e) The infrared spectrum of Q is shown below.

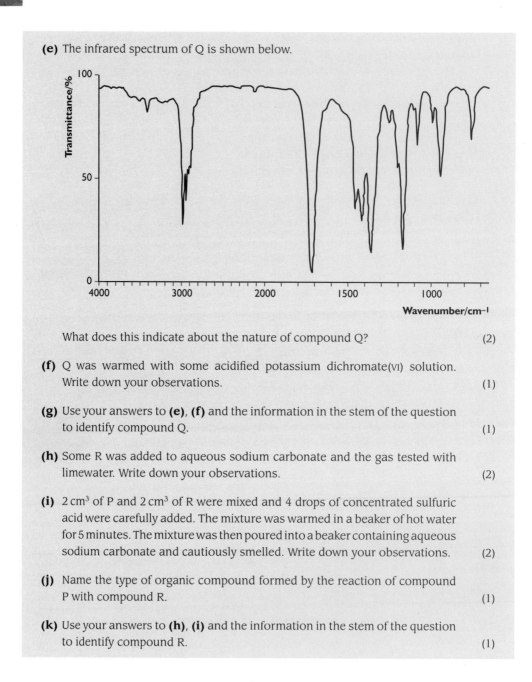

What does this indicate about the nature of compound Q? (2)

(f) Q was warmed with some acidified potassium dichromate(VI) solution. Write down your observations. (1)

(g) Use your answers to **(e)**, **(f)** and the information in the stem of the question to identify compound Q. (1)

(h) Some R was added to aqueous sodium carbonate and the gas tested with limewater. Write down your observations. (2)

(i) 2 cm³ of P and 2 cm³ of R were mixed and 4 drops of concentrated sulfuric acid were carefully added. The mixture was warmed in a beaker of hot water for 5 minutes. The mixture was then poured into a beaker containing aqueous sodium carbonate and cautiously smelled. Write down your observations. (2)

(j) Name the type of organic compound formed by the reaction of compound P with compound R. (1)

(k) Use your answers to **(h)**, **(i)** and the information in the stem of the question to identify compound R. (1)

Inorganic substances

Most of the marks are for observations. You will normally be expected to identify each unknown, and you may also be required to write an equation — e.g. for a ligand exchange reaction, a redox reaction or for a precipitation reaction.

A typical mark distribution is:

Observations	9 or 10 marks
Identification of each unknown	2 or 3 marks
Equation and/or comment	1 or 2 marks

You will be given two or three *d*-block compounds as solids or as their solutions.

Cations
The cations will be limited to Cr^{3+}, Mn^{2+}, Fe^{2+}, Fe^{3+}, Co^{2+}, Ni^{2+}, Cu^{2+} and Zn^{2+}; and possibly K^+, Na^+ and NH_4^+.

Anions
The anions will be limited to SO_4^{2-}, Cl^-, Br^-, I^-, NO_3^- and CO_3^{2-}; plus CrO_4^{2-}, $Cr_2O_7^{2-}$ and MnO_4^-.

Possible tests
Appearance of solid
You may be asked to comment on the colour and whether the solid is crystalline or a powder.

Colour	Possible ions
Colourless	Zn^{2+}
Green	Fe^{2+}, Cr^{3+}, Ni^{2+}, Cu^{2+}
Blue	Cu^{2+}
Pink	Co^{2+}, Mn^{2+}
Orange	$Cr_2O_7^{2-}$
Yellow	CrO_4^{2-}
Purple	MnO_4^-

Appearance of dilute solution

Colour	Possible ions
Colourless	Zn^{2+}, Mn^{2+}
Green	Fe^{2+}, Cr^{3+}, Ni^{2+}
Blue	Cu^{2+}
Pink	Co^{2+}
Orange	$Cr_2O_7^{2-}$
Yellow	CrO_4^{2-}
Purple	MnO_4^-

Test for a sulfate
A few drops of dilute hydrochloric acid are added to the unknown solution followed by a few drops of barium chloride solution.

Observations	Inferences
White precipitate ✓*	It is a sulfate

*Sometimes the barium chloride is added first, followed by the hydrochloric acid. The observations then are white precipitate ✓ which stays on addition of hydrochloric acid ✓.

The ionic equation for the precipitation of barium sulfate is:

$$Ba^{2+}(aq) + SO_4^{2-}(aq) \rightarrow BaSO_4(s)$$

Test for a halide

A few drops of dilute nitric acid are added to the unknown solution, followed by a few drops of silver nitrate solution.

The solubility of the precipitate formed is then tested in dilute and concentrated ammonia.

Test	Observations	Inferences
Add dilute nitric acid to a solution of the unknown and then silver nitrate* Test the precipitate: first with dilute ammonia and then (if no change) with concentrated ammonia	• White precipitate ✓ turning purple on standing ✓; soluble in dilute ammonia ✓ • Cream precipitate ✓ insoluble in dilute but soluble in concentrated ammonia ✓ • Yellow precipitate ✓ insoluble in concentrated ammonia ✓	• It is a chloride • It is a bromide • It is an iodide

*Sometimes the silver nitrate is added first, followed by the nitric acid. The observations for a chloride then are white precipitate ✓ which stays on addition of nitric acid ✓.

The ionic equation for the precipitation of a silver halide (where X stands for Cl, Br or I) is:

$$Ag^+(aq) + X^-(aq) \rightarrow AgX(s)$$

Tests for a carbonate

Test	Observations	Inferences
Heat the solid — test any gas evolved with limewater	A gas is produced which turns limewater cloudy ✓	It is a carbonate
Acid + solid — test any gas evolved with limewater	Bubbles ✓ (or fizzing or effervescence) Limewater turns milky/cloudy ✓	It is a carbonate

Test for a nitrate

Test	Observations	Inferences
Add Devarda's alloy (or aluminium powder) and dilute sodium hydroxide and warm — test any gas evolved: either with a glass rod dipped in concentrated hydrochloric acid; or with damp red litmus	• Bubbles (effervescence) ✓ • White smoke ✓ • Litmus goes blue ✓	• It is a nitrate

Addition of sodium hydroxide until in excess

Ion	Observations after a little NaOH added	Observations after excess
Fe^{2+}	Green precipitate ✓ which goes brown on standing ✓	Precipitate stays ✓
Cr^{3+}	Green precipitate ✓	Forms green solution ✓
Ni^{2+}	Green precipitate ✓	Precipitate stays ✓
Cu^{2+}	Blue precipitate ✓	Precipitate stays ✓
Mn^{2+}	Off-white (buff or sandy) ✓ precipitate which darkens on standing ✓	Precipitate stays ✓
Co^{2+}	Blue precipitate ✓ which goes pink on standing ✓	Precipitate stays ✓
Zn^{2+}	White precipitate ✓	Forms colourless solution ✓

Addition of ammonia until in excess

Ion	Observations after a little NH_3 added	Observations after excess
Fe^{2+}	Green precipitate ✓ which goes brown on standing ✓	Precipitate stays ✓
Cr^{3+}	Green precipitate ✓	Precipitate slowly dissolves in concentrated ammonia to form green solution ✓
Ni^{2+}	Green precipitate ✓	Forms blue solution ✓
Cu^{2+}	Pale blue precipitate ✓	Forms dark blue solution ✓
Mn^{2+}	Off-white (buff or sandy) ✓ precipitate which darkens on standing ✓	Precipitate stays ✓
Co^{2+}	Blue precipitate ✓ which goes pink on standing ✓	Slowly forms brown solution ✓
Zn^{2+}	White precipitate ✓	Forms colourless solution ✓

Ligand exchange with hydrochloric acid

Hydrated copper(II) and cobalt(II) ions undergo ligand exchange when a few drops of concentrated hydrochloric acid are added.

Ion	Observations
$[Cu(H_2O)_6]^{2+}$	Solution goes green ✓
$[Co(H_2O)_6]^{2+}$	Solution goes blue ✓

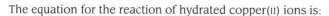

The equation for the reaction of hydrated copper(II) ions is:

$$[Cu(H_2O)_6]^{2+}(aq) + 4Cl^-(aq) \rightarrow [CuCl_4]^{2-}(aq) + 6H_2O(l)$$

Redox tests

Addition of potassium iodide
Aqueous potassium iodide is added to a solution containing the unknown ions, which oxidise the iodide ions to iodine. Starch solution is then added.

Ion	Observations	Starch added
Fe^{3+}	Red–brown solution ✓	Blue–black colour ✓
Cu^{2+}	Red–brown colour ✓ (with precipitate)	Blue–black colour ✓
$Cr_2O_7^{2-}$ or CrO_4^{2-} in acid	Brown colour ✓	Blue–black colour ✓

Addition of aqueous chlorine (chlorine water)
This is a redox reaction in which chlorine oxidises the unknown ion.

Ion	Observations
Br^-	Red colour ✓
I^-	Red–brown solution ✓; goes blue–black ✓ when starch added
Fe^{2+}	Goes slightly brown ✓; the Fe^{3+} ions can then be tested using NaOH(aq)

The ionic half-equation for the reduction of chlorine:

$$Cl_2(aq) + 2e^- \rightarrow 2Cl^-(aq)$$

Addition of other oxidising agents

Ion to which an oxidising agent is added	Observations with hydrogen peroxide as the oxidising agent	Observations with lead(IV) oxide as the oxidising agent followed by filtration
Cr^{3+} in alkaline solution	Yellow solution ✓ (of CrO_4^{2-})	Yellow solution ✓ (of CrO_4^{2-})
Fe^{2+} in acid solution	Goes slightly brown ✓; the Fe^{3+} ions can then be tested using NaOH(aq)	Goes slightly brown ✓; the Fe^{3+} ions can then be tested using NaOH(aq)

Practice example A2B3

Assume that you have been given some solid hydrated copper(II) nitrate.
Write down the observations that you would expect to make. The answers are on p. 86.

(a) Note the appearance of the solid. (1)

(b) Heat a small sample of the solid. Write down your observations. (3)

(c) Dissolve some of the solid in water and then add aqueous ammonia, with shaking, until in excess. Write down your observations. (2)

(d) Dissolve the remainder of the solid in water and add potassium iodide solution. Then add 5 drops of starch solution. Write down your observations. (2)

(e) Write an ionic equation, with state symbols, for the reaction that took place when potassium iodide was added. (1)

Practice example A2B4

Assume that you have been given some solid iron(II) sulfate.
Write down the observations that you would expect to make. The answers are on p. 86.

(a) Note the colour of the solid. (1)

(b) Dissolve the solid in water in a boiling tube. To one portion, add aqueous sodium hydroxide and allow to stand for a few minutes. Write down your observations. (2)

(c) To another portion, add dilute hydrochloric acid followed by barium chloride solution. Write down your observations. (1)

(d) To a third portion, add some solid lead(IV) oxide and some dilute sulfuric acid. Warm gently for 3 minutes and filter. Test the filtrate with aqueous sodium hydroxide. Write down your observations. (1)

Practice example A2B5

Assume that you have been given a solution of zinc bromide.
Write down the observations that you would expect to make. The answers are on p. 86.

(a) Note the appearance of the solution. (1)

(b) To one portion, add aqueous sodium hydroxide until in excess. Write down your observations. (2)

(c) To another portion, add aqueous ammonia until in excess. Write down your observations. (2)

(d) To a third portion, add dilute nitric acid followed by aqueous silver nitrate. Write down your observations. (1)

(e) To a fourth portion, add aqueous chlorine. Write down your observations. (1)

Activity C: Quantitative measurement

Full details of all the experiments assessing this activity will be given to you just before starting the assessment. It is essential that you read them carefully before starting any practical work.

Redox titrations

You will be required to titrate a solution of potassium manganate(VII), which is in a burette, with $25.0\,cm^3$ of a solution of a reducing agent to which acid has been added. No indicator is needed because the titration is stopped when you see the first permanent pink colour. This is due to a slight excess of the intensely coloured MnO_4^- ions.

The reducing agent will be supplied as a solid and it must be weighed and made up to $250\,cm^3$ in a volumetric flask.

Possible errors

- Not recording all masses to 2 decimal places.
- Not rinsing all the reducing agent solution from the beaker into the volumetric flask.
- Not shaking the volumetric flask sufficiently.
- Not rinsing out the burette with the potassium manganate(VII) solution and the pipette with the reducing agent solution.
- Not making sure that the part below the tap of the burette is filled before doing the first titration.
- Not obtaining two or more concordant titres — the difference between the largest and the smallest titres you use *to calculate the mean titre* must not be greater than $0.2\,cm^3$.
- Not recording all volumes to $0.05\,cm^3$. Thus a volume recorded as $23.6\,cm^3$ would lose a mark — it should be recorded as $23.60\,cm^3$ or $23.65\,cm^3$.

Calculations

You will be given the equation. The two most likely reactions are:

- with iron(II) compounds as the reducing agents

$$5Fe^{2+}(aq) + MnO_4^-(aq) + 8H^+(aq) \rightarrow 5Fe^{3+}(aq) + Mn^{2+}(aq) + 4H_2O(l)$$

  where the number of moles of $Fe^{2+} = 5 \times$ the moles MnO_4^-

- with ethanedioic acid, $H_2C_2O_4$, or its salts as reducing agents

$$5C_2O_4^{2-}(aq) + 2MnO_4^-(aq) + 16H^+(aq) \rightarrow 10CO_2(g) + 2Mn^{2+}(aq) + 8H_2O(l)$$

  where the number of moles of $C_2O_4^{2-} = \frac{5}{2} \times$ the moles MnO_4^-

🄴 The reaction between ethanedioate ions and manganate(VII) ions is very slow at room temperature, so the acidified ethanedioate solution must be heated before being titrated.

A typical mark scheme is:

Table of masses and titres	2 marks
Calculation of mean titre	1 mark
Accuracy of titration	7 marks
Calculation	2 or 3 marks
Comments	1 or 2 marks

Worked example A2C1

9.51 g of an iron(II) compound, Z, was weighed out, dissolved in water and made up to 250 cm³. 25.0 cm³ portions were pipetted into a conical flask and some dilute sulfuric acid added. This solution was then titrated with $0.0202 \, mol \, dm^{-3}$ potassium manganate(VII):

$$5Fe^{2+}(aq) + MnO_4^-(aq) + 8H^+(aq) \rightarrow 5Fe^{3+}(aq) + Mn^{2+}(aq) + 4H_2O(l)$$

and the mean titre was 24.05 cm³.

(a) Calculate the mass of Fe^{2+} in 250 cm³ of solution. (2)

(b) Calculate the percentage of iron in the solid Z. (1)

(c) What would be the effect on the mean titre and the % of iron if some of the compound Z had not been washed into the volumetric flask? Justify your answer. (2)

Answer

(a) Amount of MnO_4^- in mean titre = concentration × volume in dm^3

$$= 0.0202 \, mol \, dm^{-3} \times \frac{24.05}{1000} \, dm^3$$

$$= 0.0004858 \, mol$$

Amount of Fe^{2+} in 25.0 cm³ of solution $= 5 \times 0.0004858 \, mol$

$$= 0.002429 \, mol ✓$$

Amount of Fe^{2+} in 250 cm³ $= 10 \times 0.002429 \, mol$

$$= 0.02429 \, mol$$

Mass of Fe^{2+} in 250 cm³ solution = moles of $Fe^{2+} \times A_r(Fe)$

$$= 0.02429 \, mol \times 55.8 \, g \, mol^{-1}$$

$$= 1.355 \, g ✓$$

(b) % iron $= \dfrac{\text{mass of } Fe^{3+} \text{ in 250 cm}^3 \text{ solution}}{\text{mass of } Z} \times 100\%$

$$= \frac{1.355 \, g}{9.51 \, g} \times 100\%$$

$$= 14.3\% ✓$$

(c) The concentration of Fe^{2+} ions would have been lower, resulting in a smaller titre ✓ This would mean that the amount of Fe^{3+} as calculated would have been less and so the % would have been less ✓

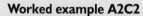

Worked example A2C2

1.59 g of a group 1 ethanedioate salt, $M_2C_2O_4$, was weighed out, dissolved in water and made up to 250 cm³. 25.0 cm³ aliquots were acidified and heated to 60°C. The hot solution was titrated with 0.0202 mol dm⁻³ potassium manganate(VII) solution. The mean titre was 23.65 cm³:

$$5C_2O_4{}^{2-}(aq) + 2MnO_4{}^-(aq) + 16H^+(aq) \rightarrow 10CO_2(g) + 2Mn^{2+}(aq) + 8H_2O(l)$$

(a) Calculate the number of moles of ethanedioate ions in 250 cm³ of the solution. (3)

(b) Use your answer to **(a)** to calculate the molar mass of $M_2C_2O_4$, and hence the identity of the group 1 metal M. (2)
[Relative atomic masses: Li = 6.9; Na = 23.0; K = 39.1, Rb = 85.5]

(c) State and explain the effect on the first titre if the burette had been wet with water, and not rinsed out with some potassium manganate(VII) solution. (1)

Answer

(a) Amount of $MnO_4{}^-$ in mean titre = concentration × volume in dm³

$$= 0.0202 \text{ mol dm}^{-1} \times \frac{23.65}{1000} \text{ dm}^3$$

$$= 0.0004777 \text{ mol } \checkmark$$

Amount of $C_2O_4{}^{2-}$ in 25.0 cm³ of solution $= \frac{5}{2} \times 0.0004777$ mol

$$= 0.001194 \text{ mol } \checkmark$$

Number of moles of $C_2O_4{}^{2-}$ in 250 cm³ of solution = 10 × 0.001194 mol

$$= 0.01194 \checkmark$$

(b) Molar mass $= \dfrac{\text{mass}}{\text{moles}}$

$$= \frac{1.59 \text{ g}}{0.01194 \text{ mol}^{-1}}$$

$$= 133 \text{ g mol}^{-1} \checkmark$$

Relative atomic mass of M × 2 = 133 − (2 × 12) − (4 × 16) = 45
Relative atomic mass of M $= \frac{1}{2} \times 45 = 22.5$
So M must be sodium (atomic mass 23) ✓

(c) The potassium manganate(VII) solution would have been more dilute, so the titre would have been larger ✓.

pH titrations

You will be asked to measure out, using a pipette, 25.0 cm³ of a solution of a weak acid and measure its pH using a probe and pH meter.

A strong alkali, such as sodium hydroxide, of known concentration is then added, initially in small amounts of between 1 and 2 cm³ to get a rough idea of the equivalence point. This is then repeated but adding the alkali in larger amounts until close to the equivalence point, when it is then added in 0.5 or 1.0 cm³ portions. After each addition, the pH of the solution is measured. The addition is continued until an excess of alkali has been added.

You will be required to plot a graph of pH against volume of alkali, and then perform some calculations and make comments about accuracy.

Errors in technique

- As with all titrations, make sure that you rinse out the pipette and burette with the appropriate solutions.
- Make sure that the part of the burette below the tap is filled before starting the pH titration.
- Make sure that you add the alkali in small amounts to start with and close to the end point. In between these it can be added in 5 or 10 cm³ portions.
- Make sure that you stir the solution thoroughly with the pH probe after each addition of alkali. Failure to do this properly is a common error.

Errors in recording

- Make sure that you record all volumes to 0.05 cm³ and all pH values to 1 or 2 decimal places.

Graph plotting

- Make sure that you use at least half the graph paper — do not draw a tiny graph at the bottom left of the graph paper.
- Make sure that you label the y-axis (pH) and the x-axis (volume of alkali/cm³).
- The pH value before any alkali is added should be about pH = 3. It should rise sharply over the next 2 to 3 cm³ and then flatten off, slowly increasing before rising almost vertically as the equivalence point is reached (when the acid and the alkali are in a 1 : 1 ratio by moles). It then flattens off again after about 5 cm³ of excess alkali has been added.
  To see what this looks like, look at the graph in worked example A2C3.
  The buffer region runs from after the initial rise in pH (after about 5 cm³ alkali has been added) to about 5 cm³ before the almost vertical part. This is when the ratio of [weak acid] : [salt] varies from about 5 : 1 to 1 : 5.

A typical mark scheme is:

Table of readings	4 or 5 marks
Graph	2 or 3 marks
Accuracy	2 or 3 marks
Calculations from graph	3 marks
Comments	1 or 2 marks

Worked example A2C3

25.0 cm³ of a solution of a weak acid HX was measured into a beaker and its pH measured. A burette was filled with 0.125 mol dm⁻³ sodium hydroxide solution. This was added a little at a time, with stirring, measuring the pH after each addition. The results are shown overleaf.

Volume of alkali/cm³	pH	Volume of alkali/cm³	pH
0	2.9	19.00	5.7
1.00	3.8	19.50	5.9
2.55	4.3	20.00	8.8
5.05	4.6	21.05	11.4
7.50	4.7	23.00	11.7
12.50	5.0	25.00	12.0
15.05	5.2	30.05	12.3
17.50	5.5	40.00	12.7

(a) Plot a graph of pH on the y-axis against volume of sodium hydroxide on the x-axis. (6)

(b) Use your graph to find the volume of alkali needed to react with the 25.0 cm³ of the acid, HX (2) and the pH at the equivalence point (1).

(c) Use your answer to **(b)** and the concentration of the sodium hydroxide solution given to calculate the concentration of the acid, HX. (2)

(d) Measure the pH at the point when half the acid has been neutralised, and hence calculate the value of pK_a of the acid HX. (2)

(e) Use your answer to **(d)** to calculate the value of K_a of the acid. (1)

(f) Over what pH range would the solution in the pH titration be able to act as a buffer? (1)

(g) Comment on the accuracy of this method for finding the concentration of the acid compared with a normal acid/base titration. (1)

Answer

(a)

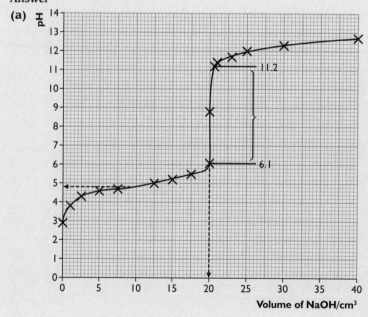

The scoring for table and this graph would be:

- small additions of alkali at start and close to equivalence point ✓ and larger additions at other stages ✓
- All volumes recorded to $\pm 0.05\,cm^3$ and all pH values to at least 1 decimal place ✓
- pH always rising as alkali added ✓
- sensible scale and axes labelled with points properly plotted ✓
- steep rise in pH over first few cm^3, then an s-shaped curve with an almost vertical rise near $20\,cm^3$ followed by flattening off towards pH = 12.7 ✓

(b) $20.0\,cm^3$ (value from 19 to 21 ✓✓; from 18 to 19, or 20 to 21 ✓)

pH = 8.7 (value from 8.2 to 9.3) ✓

[These are the volume and pH at halfway up the almost vertical part]

(c) Amount of alkali = $0.125\,mol\,dm^{-3} \times \dfrac{20.0}{1000}\,dm^3$

$$= 0.00250\,mol$$

And this is the number of moles of HX ✓

Concentration of HX = $\dfrac{\text{moles of HX}}{\text{volume}}$

$$= 0.100\,mol\,dm^{-3} ✓$$

(d) pH at $10\,cm^3$ added = 4.8 ✓, hence pK_a = 4.8 ✓

🖉 This is because when half of the acid has been neutralised the concentration of the acid is the same as the concentration of the salt. But $K_a = \dfrac{[H^+][\text{salt}]}{[\text{acid}]}$ and $\dfrac{[\text{salt}]}{[\text{acid}]} = 1$.

So $K_a = [H^+]$, and pK_a = pH.

(e) $K_a = 10^{-pK_a} = 10^{-4.8}$

$$= 1.6 \times 10^{-5}\,(mol\,dm^{-3}) ✓$$

(f) Between pH 4.5 and 5.2 ✓

(g) It is not nearly as accurate because it is difficult to read the volume at halfway up the vertical part with an accuracy better than $\pm 0.5\,cm^3$ ✓.

Kinetic experiments

Following a reaction at constant temperature

This type of experiment is performed to find the order of reaction of one of the reactants.

Full details will be given and care must be taken to follow these instructions carefully. You must read them through completely before starting the experiment, because once you have started there will not be time to work out what to do next.

The procedure will be:

(1) Measure out the reagents and catalyst solutions in two separate beakers.

(2) Prepare the quenching reagent in a conical flask.

(3) Fill a burette with a solution that will react rapidly with one reagent (or product).

(4) Mix the contents of the two beakers and start a clock.

(5) After 5 minutes remove a 10 cm³ sample, quench the reaction and titrate against the solution in the burette.

(6) Remove further samples every 5 minutes and quench and titrate.

This method can be used to find the order of the reaction between propanone and iodine in the presence of an acid catalyst:

$$CH_3COCH_3(l) + I_2(aq) \xrightarrow{\text{H}^+ \text{ catalyst}} CH_3COCH_2I(aq) + HI(aq)$$

The concentrations of propanone and acid must be much larger than that of the iodine solution. The iodine in the samples removed is titrated against sodium thiosulfate solution.

A typical mark scheme is:

Table of volumes of reagents, titres and times	3 or 4 marks
Plotting of graph	2 or 3 marks
Interpretation of graph	2 or 3 marks
Accuracy of experiment	2 marks
Comments on experiment	3 or 4 marks

Worked example A2C4

Using a measuring cylinder, 25 cm³ of a solution of propanone was mixed with 25 cm³ of dilute sulfuric acid in a beaker. 25 cm³ of a dilute iodine solution was measured out in another measuring cylinder.

Using a third measuring cylinder, 20 cm³ portions of sodium hydrogencarbonate solution were put into conical flasks.

A burette was filled with 0.0500 mol dm⁻³ sodium thiosulfate solution.

The iodine solution was poured into the beaker containing the propanone and acid, stirred and a clock started. After 5 minutes, 10 cm³ of the reaction mixture was removed and run into the sodium hydrogencarbonate to quench the reaction. This was then immediately titrated with the sodium thiosulfate solution.

Another 10 cm³ portion was then removed, quenched and titrated, until five portions had been removed. The results are shown below.

Volume of propanone solution/cm³	Volume of sulfuric acid/cm³	Volume of iodine solution/cm³	Volume of sodium hydrogencarbonate solution/cm³
25	25	25	20

Time/min	5	10	15	20	25
Titre/cm³	18.0	14.4	10.8	7.6	4.0

(a) Plot a graph of titre on the *y*-axis against time on the *x*-axis. (8)

(b) Calculate the slope (gradient) of this graph. (1)

(c) What does the slope of the graph measure? (1)

(d) Use your answer to **(c)** to find the order of the reaction with respect to iodine. Justify your answer. (3)

(e) In this experiment, the concentration of propanone was much larger than that of the iodine. Why is this necessary? (1)

(f) Why is it not necessary to use a pipette to measure the volume of propanone, sulfuric acid and iodine solutions? (1)

(g) Why is it necessary to add the reaction mixture to aqueous sodium hydrogencarbonate? (3)

(h) What would be produced if sodium hydroxide and not sodium hydrogencarbonate had been used to quench the reaction? (1)

(i) How would you modify the procedure to enable the order with respect to propanone be found? (2)

Answers

(a)

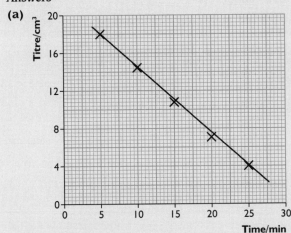

The scoring would be:
- Table 1: all volumes of initial solutions ✓
- Table 2: all titres to 1 decimal place ✓; and all times to nearest minute ✓; titres decreasing ✓
- Axes labelled and reasonable scale ✓; points plotted correctly and best-fit straight line drawn ✓
- Accuracy of graph ✓✓

(b) Gradient $= \dfrac{(18 - 4)\,\text{cm}^3}{(25 - 5)\,\text{min}}$

$\qquad = 0.70\,\text{cm}^3\,\text{min}^{-1}$ ✓

(c) It is proportional to the rate of the reaction ✓

(d) The volume of thiosulfate is proportional to the concentration of iodine in the sample removed ✓; it is zero order ✓ because the rate of reaction is constant ✓

(e) So that the concentration of propanone does not change significantly during the experiment ✓

(f) Because you do not need to know the exact amounts of each as the order of reaction must be a whole number or zero ✓

(g) The sodium hydrogencarbonate neutralises the acid catalyst ✓; and so stops the reaction ✓; otherwise iodine would be used up by the propanone during the titration ✓

(h) The mixture of propanone, iodine and strong alkali would produce a precipitate of triiodomethane (iodoform) ✓

(i) Double the concentration of propanone and see how the slope alters ✓; if it also doubles, the reaction is first order with respect to propanone ✓

Finding the activation energy of a reaction

This method requires the time, t, for a particular stage in the reaction to be measured at a certain temperature, T. The experiment is then repeated at different temperatures. The Arrhenius equation connects the activation energy to the rate constant of the reaction:

$$\log_{10} k = \frac{-E_a}{2.3R} \times \frac{1}{T}$$

where R is the gas constant and T is the temperature in kelvin.

A graph of $\log_{10} k$ against $\frac{1}{T}$ will give a straight line of slope $\frac{-E_a}{2.3R}$. However, the rate constant is difficult to measure and a graph of log(rate), or the log of anything proportional to the rate such as $\frac{1}{t}$, will also give a straight line of slope $\frac{-E_a}{2.3R}$.

If a graph of $\log_{10}\frac{1}{t}$ is plotted against $\frac{1}{T}$, the slope is $\frac{-E_a}{2.3R}$ (see worked example A2C5).

A typical mark scheme is:

Table of readings + calculation of $\log\frac{1}{t}$ and $\frac{1}{T}$	5 or 6 marks
Plotting graph and interpretation	3 or 4 marks
Calculation and accuracy of activation energy	4 marks
Comments	1 or 2 marks

Activation energy of the reaction between sodium thiosulfate and acid

Sodium thiosulfate reacts with nitric acid to produce a precipitate of sulfur:

$$S_2O_3^{2-}(aq) + 2H^+(aq) \rightarrow S(s) + SO_2(aq) + H_2O(l)$$

The experiment is designed to time how long it takes for enough sulfur to be produced to hide a cross on the reaction vessel when looked at through the reaction mixture. The experiment is repeated at different temperatures.

Worked example A2C5

A cross was marked on the outside of a boiling tube with a black waterproof marker. $10\,cm^3$ of a solution of sodium thiosulfate was put in the boiling tube and its temperature measured. $10\,cm^3$ of dilute nitric acid was put in a test tube. This was then poured into the boiling tube and the solution stirred with a thermometer. The time taken for the cross to become invisible was recorded and the temperature of the solution retaken.

More sodium thiosulfate and nitric acid was then warmed separately in a beaker of water and the experiment repeated.

The results are shown below.

Time, t/s	Initial temperature/°C	Final temperature/°C	Mean temperature/°C
120	20	22	21
54	35	35	35
26	50	48	49
15	62	58	60

$\frac{1}{t}$ /s^{-1}	$\log_{10}\frac{1}{t}$	Mean temperature, T/K	$\frac{1}{T}$ /K^{-1}
0.00833	-2.08	294	0.00340
0.0185	-1.73	308	0.00325
0.0385	-1.42	322	0.00311
0.0667	-1.18	333	0.00300

(a) Plot a graph of $\log_{10}\frac{1}{time}$ against $\frac{1}{kelvin\ temperature}$. (3)

✐ The graph should have -1.0 at the top going down to -2.2 on the y-axis and 0.00300 on the left going up to 0.00340 on the right on the x-axis

(b) What does the shape of the graph tell you about the relationship between the rate of reaction and the temperature? (1)

(c) Use your graph to calculate the slope of the line you have drawn. (1)

(d) Calculate the activation energy of the reaction. Give a sign and units with your answer. (3)

(e) Why did the temperature rise slightly in the first experiment? (1)

Answers

(a)

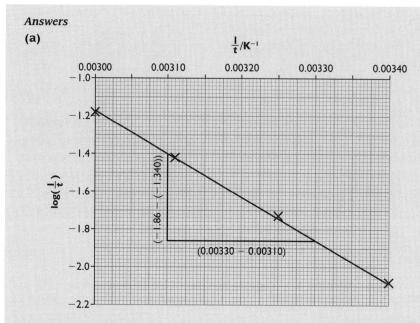

(b) Because $\dfrac{1}{\text{time}}$ is proportional to the rate, the log of the rate is proportional to $\dfrac{1}{T}$ ✓

(c) Slope $= \dfrac{-1.86 - (-1.40)}{(0.00330 - 0.00310)\text{K}^{-1}}$

$= \dfrac{-0.46}{0.0002\ \text{K}^{-1}}$

$= -2300\ \text{K}$ ✓

(d) $E_a = -\text{slope} \times 8.31\,\text{J}\,\text{K}^{-1}\,\text{mol}^{-1} \times 2.3$

$= -(-2300\,\text{K} \times 8.31\,\text{J}\,\text{K}^{-1}\,\text{mol}^{-1} \times 2.3)$

$= +43960\,\text{J}\,\text{mol}^{-1}$

$= +44\,\text{kJ}\,\text{mol}^{-1}$ ✓

Value in range 38–50 ✓✓; within range 33–55 ✓

(e) Because the reaction is exothermic ✓.

Activation energy of the reaction between magnesium and acid

This is another experiment that is suitable for measuring the time taken for a certain point to be reached and for repeating at different temperatures.

If a large excess of acid is used, the rate of reaction will be approximately constant because the surface area of the magnesium ribbon does not alter significantly. Thus the rate is proportional to $\dfrac{1}{\text{time}}$, where the time is that for the magnesium ribbon to disappear (see Test question TQ6 on p. 84).

Activity D: Preparation

In all preparations, you will be asked to work out the percentage yield. This will score 3 of the 12 marks in the assessment.

The method is:

(1) Work out the molar masses of both the reactant that you measured out and the product.

(2) Use the expression number of moles $= \dfrac{\text{mass}}{\text{molar mass}}$ to calculate the number of moles of reactant.

(3) If the reactant and product are in a $1:1$ ratio in the balanced equation for the reaction, the **theoretical** number of moles of product = the number of moles of reactant.

(4) The theoretical mass of the product = moles × molar mass.

(5) % yield $= \dfrac{\text{actual mass of product}}{\text{theoretical mass of product}} \times 100\%$

The percentage yield is **not** $\dfrac{\text{mass of product}}{\text{mass of reactant}} \times 100\%$.

Organic preparations

You will be asked to prepare an organic compound and then purify it. Finally its melting or boiling point will be measured.

Full details of the procedure will be given and you should follow them carefully.

Technique

- You must know how to set up apparatus for simple distillation (see p. 42) and for heating under reflux, as shown opposite.

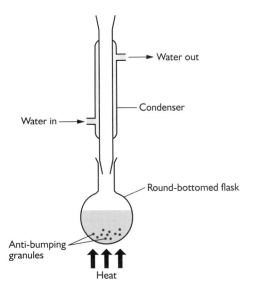

Water out

Condenser

Water in

Round-bottomed flask

Anti-bumping granules

Heat

- Make sure that the water flows into the condenser at the bottom and out at the top.
- Check that all the joints are firm and that they will not leak.
- When carrying out distillation, check that the thermometer is opposite the entrance of the condenser.
- You must know how to recrystallise a solid.
- You must know how to measure the melting point of a solid.

Recording

All masses must be given to 2 decimal places and volumes to 1 cm³.

Melting and boiling temperatures should be recorded to 1°C.

Methods of measuring melting temperatures

There are two common procedures.

(1) This method is only to be used if you have sufficient amounts of the solid. Some of the purified solid is placed in a boiling or test tube.

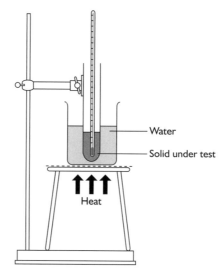

The water or oil is heated until the solid begins to melt. Note this temperature.

Now remove the tube from the beaker and allow it to cool while stirring the molten solid with the thermometer. Note the temperature at which solid first begins to form. The melting temperature is the mean of these two values.

(2) This method should be used when small amounts of solid are available. Take a capillary tube that has been sealed at one end and fill it to a depth of 0.5 cm with your purified solid.

Attach it, using a rubber band, to a thermometer and insert the thermometer in a water or an oil bath.

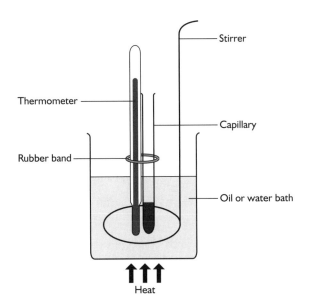

Heat the bath quickly until its temperature reaches about 20°C below the expected melting point. Then gradually increase the temperature until the solid melts. Note this approximate melting temperature.

Let the oil or water bath cool slightly, and repeat with another sample of the solid in another capillary tube. Heat gently up to close to the approximate melting temperature of the solid, as found earlier. Record the temperature at which it melts.

Marking

The mark scheme for organic preparations will be similar to the following:

Table of masses and/or volumes	2 marks
Calculation of theoretical % yield and actual % yield	3 marks
% yield	>50% for 2 marks >10% but <50% for 1 mark
Description of product	1 mark
Value of melting or boiling temperatures	±2° = 2 marks ±5° = 1 mark
Comments	2 marks

Preparation of aspirin

Aspirin is a solid phenolic ester made by the reaction of 2-hydroxybenzoic acid (salicylic acid) and ethanoic anhydride in the presence of phosphoric acid catalyst:

Salicylic acid $\quad\quad + \quad (CH_3CO)_2O \quad\longrightarrow\quad$ Aspirin $\quad + \quad CH_3COOH$

Some 2-hydroxybenzoic acid is weighed out and put in a distillation flask together with excess ethanoic anhydride. Then 5 drops of phosphoric acid are added and a reflux condenser is fitted. The mixture is heated on a boiling water bath for 5 minutes and then some water is carefully added down the reflux condenser. This hydrolyses the excess ethanoic anhydride.

The mixture is poured into a beaker of cold water and stood in an ice bath. The mixture is filtered using suction filtration, and the solid obtained is recrystallised using hot water.

The pure, dry aspirin is weighed, and finally the melting temperature of the aspirin is measured. This should be between 134 and 135°C.

Worked example A2D1

The following questions are about the preparation of aspirin.

(a) Describe the appearance of your sample of aspirin. (1)

(b) Calculate the theoretical yield of aspirin starting with 4.00 g of 2-hydroxybenzoic acid. (2)

(c) Why is the actual yield less than 100%? (1)

(d) Why is it necessary to heat the sample slowly (at about 2°C per minute close to the melting point) when measuring the melting temperature? (1)

(e) What type of impurity is removed in the first filtration during recrystallisation? (1)

(f) What type of impurity is removed in the second filtration during recrystallisation? (1)

(g) What would be formed if 2-hydroxybenzoic acid was warmed with methanol in the presence of concentrated sulfuric acid? (1)

Answers

(a) It is a white slightly crystalline solid ✓

(b) Amount of 2-hydroxybenzoic acid $= \dfrac{\text{mass}}{\text{molar mass}}$

$$= \dfrac{4.00\,g}{138\,g\,mol^{-1}}$$

$$= 0.0290\,mol \checkmark$$

Theoretical amount of aspirin $= 0.0290\,mol$

Theoretical yield $=$ number of moles $\times$ molar mass

$$= 0.0290\,mol \times 180\,g\,mol^{-1}$$

$$= 5.22\,g \checkmark$$

> ℯ Your mass of purified aspirin should be similar to the mass of 2-hydroxybenzoic acid originally taken.
>
> **(c)** Either because the reaction was not complete or because some aspirin was left dissolved in the water during purification ✓
>
> **(d)** So that the temperature measured and that of the solid in the tube are the same ✓
>
> **(e)** Impurities that are insoluble in hot water ✓
>
> **(f)** Impurities that are soluble in cold water ✓
>
> **(g)** Methyl 2-hydroxybenzoate (oil of wintergreen) ✓

Preparation of methyl 3-nitrobenzoate

Methyl 3-nitrobenzoate is a solid ester made by the nitration of methyl benzoate.

A known volume of methyl benzoate is carefully mixed with some concentrated sulfuric acid in a flask. A nitrating mixture of cooled concentrated sulfuric and nitric acid is then added in small portions and the temperature maintained between 5 and 15°C. The flask is allowed to warm up to room temperature and left to stand for 15 minutes. The contents are then poured onto some ice and stirred. The mixture is filtered using suction, and the solid washed with a little ice-cold ethanol.

The solid is then recrystallised using ethanol as the solvent and is allowed to dry. When dry, it is weighed and its melting point is determined in the usual way. The melting point should be between 78 and 80°C.

Worked example A2D2

The following questions are about the preparation of methyl 3-nitrobenzoate.

(a) Describe the appearance of your product. (1)

(b) Calculate the theoretical yield of methyl 3-nitrobenzoate starting with 5.00 cm³ of methyl benzoate (density = 1.09 g cm⁻³). (3)

(c) If the mass of your product was 5.05 g, what was your % yield? (2)

(d) Why must the temperature during the nitration be kept at or below room temperature? (1)

(e) Identify the electrophile in this reaction. (1)

(f) Why is the recrystallised solid washed with a little ice-cold ethanol before being allowed to dry? (1)

Answers

(a) It is a white solid ✓

(b) Mass of methyl benzoate = volume × density

$$= 5.00 \, cm^3 \times 1.09 \, g \, cm^{-3}$$
$$= 5.45 \, g \checkmark$$

$$\text{Amount of methyl benzoate} = \frac{mass}{molar \ mass}$$
$$= \frac{5.45 \, g}{136 \, g \, mol^{-1}}$$
$$= 0.0401 \, mol \checkmark$$

Theoretical number of moles of methyl 3-nitrobenzoate = 0.0401 mol

Theoretical mass of methyl 3-nitrobenzoate = moles × molar mass

$$= 0.0401 \, mol \times 181 \, g \, mol^{-1}$$
$$= 7.25 \, g \checkmark$$

(c) % yield $= \dfrac{5.05 \, g}{7.25 \, g} \times 100\%$

$$= 70\% \checkmark; \text{ plus } \checkmark \text{ for } >50\%$$

(d) If allowed to rise, a second nitro group would be substituted into the ring ✓

(e) The nitronium ion, NO_2^+ ✓

(f) To remove any solution that will contain soluble impurities ✓

Inorganic preparation

You will be given full instructions which must be followed carefully.

You will be required to prepare a solid containing a complex transition metal ion, such as $[Cu(NH_3)_4]^{2+}SO_4^{2-}$ or chromium(II) ethanoate, $[Cr_2(CH_3COO)_4(H_2O)_2]$.

Preparation of tetraamminecopper(II) sulfate

Some solid copper(II) sulfate, $CuSO_4.5H_2O$, is weighed out and dissolved in the minimum of water. Aqueous ammonia solution is added steadily, with stirring, until in excess.

Ethanol is added to this solution and a precipitate of the copper(II) complex salt $[Cu(NH_3)_4]SO_4$ is formed. This is separated from the aqueous layer by suction filtration and the solid washed with ethanol and allowed to dry. Finally it is weighed.

A typical mark scheme for an inorganic preparation is:

Table of masses	2 or 3 marks
Observations during the preparation	1 or 2 marks
Appearance of product	1 or 2 marks
Calculation of % yield	3 marks
Value of yield	1 or 2 marks
Comments	2 or 3 marks

Errors in technique
- Dissolving the solid in too much water.
- Leaving behind some of the solid product before suction filtration.
- Washing the filtered solid with too much solvent.

Errors in recording
- Not making sure that all masses are recorded to 2 decimal places — a mass written as 12.6 will lose a mark.
- Not checking all subtractions when working out the mass of solid taken and the mass of product.

Worked example A2D3

The following questions assume that $[Cu(NH_3)_4]SO_4$ has been prepared.

(a) What would you observe as the ammonia solution was added to the copper(II) sulfate solution? (2)

(b) Describe the appearance of your product. (1)

(c) Assume that you took 6.00 g of solid hydrated copper(II) sulfate, $CuSO_4.5H_2O$, calculate the theoretical yield of the complex salt, $[Cu(NH_3)_4]SO_4$. (2)

(d) Assume that you made 4.68 g of product, what is your yield? (2)

(e) Name the type of reaction between hydrated copper(II) ions and ammonia. (1)

(f) Suggest a reason why the yield is less than 100%. (1)

(g) What was the purpose of adding ethanol to the reaction mixture? (1)

(h) Why was the product washed with ethanol after suction filtration? (1)

Answers
(a) The blue solution would form a pale blue precipitate ✓; which dissolves to give a dark blue solution ✓ with excess ammonia

(b) It is a dark blue–purple solid ✓; with needle shaped crystals ✓

(c) Amount of hydrated copper sulfate $= \dfrac{\text{mass}}{\text{molar mass}}$

$$= \frac{6.00\,\text{g}}{249.6\,\text{g mol}^{-1}}$$

$$= 0.0240\,\text{mol} ✓$$

Theoretical amount of $[Cu(NH_3)_4]SO_4 = 0.0240\,\text{mol}$

Theoretical mass of $[Cu(NH_3)_4]SO_4 = \text{moles} \times \text{molar mass}$

$$= 0.0240\,\text{mol} \times 227.6\,\text{g mol}^{-1}$$

$$= 5.47\,\text{g} ✓$$

(d) % yield $= \dfrac{4.68\,\text{g}}{5.47\,\text{g}} \times 100\%$

$$= 86\% ✓;\ \text{plus} ✓ \text{ for this value being} >25\%$$

(e) Ligand exchange ✓

(f) Some of the product stayed dissolved in the aqueous solution ✓

(g) To dissolve the water causing the complex to precipitate ✓

(h) To remove dissolved impurities ✓

Preparation of a chromium(II) ethanoate complex

Weigh some sodium dichromate(VI), dissolve it in water and pour the solution into a round-bottomed flask. Add a mixture of powdered and granulated zinc. Pour some saturated sodium ethanoate solution in a boiling tube and assemble the apparatus as in the diagram below. Carefully add a mixture of concentrated hydrochloric acid and water made up in a 2 : 1 ratio. Make sure that the screw cap is loose.

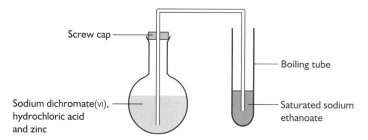

The dichromate(VI) ions are first reduced to green Cr^{3+} ions, and then to blue Cr^{2+} ions.

When the solution is blue, screw the cap shut. The solution of Cr^{2+} ions is forced out into the sodium ethanoate solution by the pressure of the hydrogen being produced. A red precipitate of the neutral chromium(II) complex $[Cr_2(CH_3COO)_4(H_2O)_2]$ is formed.

Filter this precipitate under suction, wash it with a trace of ice-cold propanone and weigh it.

Worked example A2D4

(a) Calculate the theoretical yield of the chromium(II) ethanoate complex — assume that you took 2.22 g of sodium dichromate(VI) (2)

(b) Calculate the % yield of the experiment — assume that you made 1.91 g of the complex. (2)

(c) Why is the yield less than 100%? (1)

(d) Write the ionic half-equations for the reduction of $Cr_2O_7^{2-}$ ions in acid solution to Cr^{3+} ions, and Cr^{3+} ions to Cr^{2+} ions. (2)

(e) Why was the screw cap left open until the solution went blue? (1)

(f) Why do the chromium(II) ethanoate complex and the hydrated chromium(II) ion have different colours? (1)

Answers

(a) Amount of sodium dichromate(VI) $= \dfrac{\text{mass}}{\text{molar mass}}$

$$= \dfrac{2.22\,\text{g}}{262\,\text{g mol}^{-1}}$$

$$= 0.00847\,\text{mol}$$

Theoretical amount of chromium(II) ethanoate $= 0.00847\,\text{mol}$ ✓
Theoretical mass of the chromium(II) ethanoate $=$ moles $\times$ molar mass

$$= 0.00847\,\text{mol} \times 376\,\text{g mol}^{-1}$$

$$= 3.19\,\text{g}\ ✓$$

(b) % yield $= \dfrac{1.91\,\text{g}}{3.19\,\text{g}} \times 100\%$

$$= 60\%\ ✓;\ \text{plus}\ ✓\ \text{for value} >25\%$$

(c) Either some solution of the Cr^{2+} ions remained in the round-bottomed flask; or some chromium(II) ethanoate complex stayed dissolved in the water ✓

(d) $Cr_2O_7{}^{2-}(aq) + 14H^+(aq) + 6e^- \rightarrow 2Cr^{3+}(aq) + 7H_2O(l)$ ✓
$Cr^{3+}(aq) + e^- \rightarrow Cr^{2+}(aq)$ ✓

(e) Because hydrogen is produced by the reaction of zinc and acid and pressure must not be allowed to build up until all the chromium has been reduced to Cr^{2+} ✓

(f) The different ligands split the d-orbitals of the Cr^{2+} ion to a different extent ✓

content guidance

Questions
&
Answers

This section contains examples of the type of experiments and questions that you will encounter in Activities B, C and D in your internally assessed practical in Units 3 and 6. It also contains the answers to these Test questions and to the Practice examples. A tick (✓) represents a scoring point.

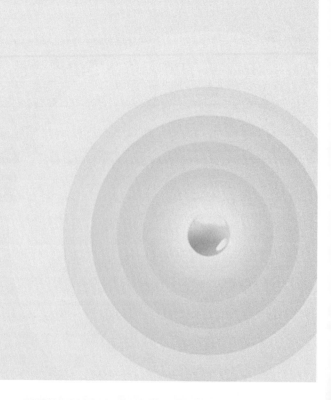

Test questions

AS questions

Ask your teacher if you can do these experiments. The identity of the unknowns and the concentrations of solutions are given in the answers section.

You must ask your teacher to carry out a risk assessment before doing these experiments.

Test question 1

Carry out the following tests on the two solid inorganic salts, A and B, and a solution of an inorganic salt, C.

(a) **Carry out a flame test on solids A and B. Write down your observations with A and with B.** **(2)**

(b) **Make a solution of some of A in 3 cm³ of water and add 4 drops of dilute hydrochloric acid, followed by 5 drops barium chloride solution. Write down your observations.** **(1)**

(c) **Using your observations in (a) and (b), identify A by writing its formula.** **(1)**

(d) **Write the ionic equation for the reaction observed in (b).** **(1)**

(e) **Heat some of solid B in a test tube and pass the gas evolved into limewater. Write down your observations.** **(2)**

(f) **Heat 3 cm³ of water in a test tube — almost to boiling. Add some solid B and test any gas evolved with limewater. Write down your observations.** **(2)**

(g) **Using your observations in (a), (e) and (f), write the formula of B.** **(1)**

(h) **To 2 cm³ of the solution of C, add 4 drops of dilute nitric acid, followed by 5 drops of aqueous silver nitrate. Then add aqueous ammonia until there is no further change. Write down your observations.** **(2)**

(i) **To a second 2 cm³ portion of aqueous C, add 10 drops of dilute sodium hydroxide solution. Write down your observations.** **(1)**

(j) **Using your observations in (h) and (i), write the formula of C.** **(1)**

Total: 14 marks

Test question 2

Carry out the following tests on three organic compounds, P, Q and R.

(a) **Mix 1 cm³ aqueous potassium dichromate(VI) and dilute sulfuric acid in a test tube, and add 5 drops of P. Place the test tube in a beaker of warm water for a few minutes. Write down your observations.** **(1)**

(b) Working in a fume cupboard, put $1\,cm^3$ of **P** in a dry test tube. Then carefully add half a spatula of phosphorus(V) chloride. Test the gas evolved with damp blue litmus paper. Write down your observations. (2)

(c) Look at the mass spectrum of **P** below.

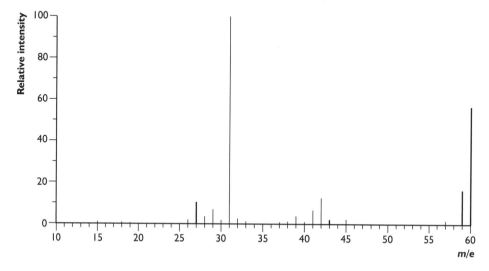

By considering your answers to (a) and (b) along with the m/e value of the molecular ion and the fragment ion at $m/e = 31$ in the mass spectrum, write the formula of **P**. (1)

(d) Pour $3\,cm^3$ of aqueous silver nitrate into a test tube and add 4 drops of **Q**. Then add aqueous ammonia until there is no further change. Write down your observations. (2)

(e) The mass spectrum of **Q** has two molecular ion peaks of equal intensity at m/e values of 136 and 138. Using this and your answer to (d), and the fact that **Q** has a branched chain, write a formula of **Q**. (1)

(f) Put a few drops of **R** on a crucible lid and ignite it. Write down your observations. (1)

(g) Pour $2\,cm^3$ of bromine water into a test tube and add 4 drops of **R**. Gently shake the test tube. Write down your observations. (2)

(h) Using your answers to (f) and (g), state the functional group in **R** and suggest another feature of the molecule. (2)

(i) The infrared spectrum below is that of **P**, **Q** or **R**. Examine the spectrum and, using your data booklet, identify the compound which has this spectrum. Justify your choice. (2)

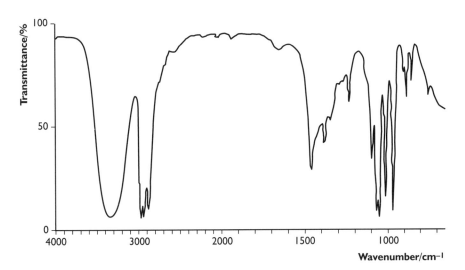

Total: 14 marks

Test question 3

You are supplied with:
- a sample of a solid iron(III) compound labelled **Z**
- aqueous $0.100\,\text{mol dm}^{-3}$ sodium thiosulfate
- aqueous $0.5\,\text{mol dm}^{-3}$ potassium iodide
- dilute sulfuric acid
- starch solution.

Iron(III) ions react with iodide ions according to the equation:

$$2Fe^{3+}(aq) + 2I^-(aq) \rightarrow 2Fe^{2+}(aq) + I_2(aq)$$

The liberated iodine can be titrated with aqueous sodium thiosulfate:

$$I_2(aq) + 2S_2O_3{}^{2-}(aq) \rightarrow 2I^-(aq) + S_4O_6{}^{2-}(aq)$$

(1) Weigh the weighing bottle containing **Z**. Copy Table 1 below and record the mass of the bottle with **Z**.

(2) Tip the solid **Z** into a $150\,\text{cm}^3$ beaker and reweigh the bottle. Record the mass of the empty bottle in the table. Then dissolve **Z** in water, transfer the solution into a $250\,\text{cm}^3$ volumetric flask and make it up to the mark with distilled water. Shake the flask thoroughly.

(3) Rinse out and then fill a burette with the sodium thiosulfate solution.

(4) Rinse out a pipette with solution **Z**, then use it to transfer $25.0\,\text{cm}^3$ aliquots into a conical flask.

(5) Use measuring cylinders to pour $10\,\text{cm}^3$ of dilute sulfuric acid and $10\,\text{cm}^3$ of potassium iodide solution into the conical flask.

(6) Titrate the solution in the conical flask with the sodium thiosulfate solution until the colour has faded to pale yellow. Then add 10 drops of starch indicator and continue to add the sodium thiosulfate drop by drop until the blue–black colour has disappeared. Record your burette readings and the titres in a table (see Table 2).

(7) Repeat steps 4–6 until you obtain two consistent titres.

Table 1

Mass of weighing bottle with Z/g	
Mass of emptied weighing bottle/g	
Mass of Z/g	

Table 2

Titration number	1	2	3	4
Burette reading (final)/cm³				
Burette reading (initial)/cm³				
Titre/cm³				

List the titres you will use to calculate the mean titre.
Write down the mean titre (cm³ of 0.100 mol dm⁻³ sodium thiosulfate). **(10)**

(a) Calculate the amount (in moles) of sodium thiosulfate in the mean titre. **(1)**

(b) Using the equations given earlier and your answer to (a), calculate the amount (in moles) of Fe^{3+} ions in 25.0 cm³ of solution, and hence in 250 cm³ of solution. **(1)**

(c) Using your answer to (b), calculate the mass of iron in 250 cm³ of solution. ($A_r(Fe) = 55.8$) **(1)**

(d) Using your answer to (c) and the mass of Z taken, calculate the percentage of iron in compound Z. **(1)**

Total: 14 marks

A2 questions

Ask your teacher if you can do these experiments. The identity of the unknowns and the concentrations of solutions are given in the answers section.

You must ask your teacher to carry out a risk assessment before doing the experiments.

Test question 4

You are supplied with three organic compounds **X**, **Y** and **Z**, each containing four carbon atoms. None of the compounds has a branched carbon chain.

(a) **Look at the spectrum of X below and comment on the information that this provides.** (2)

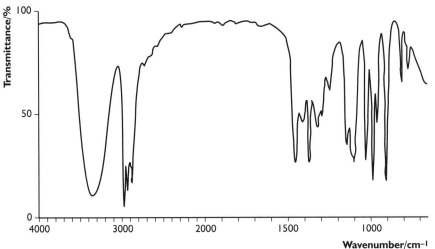

(b) **Make up a solution of equal parts of aqueous potassium dichromate(VI) and dilute sulfuric acid. Divide this into three in three separate test tubes. Add 4 drops of X to one, 4 drops of Y to another and 4 drops of Z to the third. Warm each in a beaker of hot water for 3 minutes. Write down your observations with X, Y and Z.** (3)

(c) **Add 5 drops of 2,4-dinitrophenylhydrazine solution in turn to X, to Y and to Z. Write down your observations with X, Y and Z.** (3)

(d) **Make up a solution by adding aqueous iodine to about $10\,cm^3$ of dilute sodium hydroxide until there is a faint brown colour. Pour equal amounts into three test tubes. Add 4 drops of each of X, of Y and of Z to the separate test tubes and allow to stand for 3 minutes. Write down your observations with X, Y and Z.** (3)

(e) **Identify X, Y and Z by name or formula.** (3)

Total: 14 marks

Test question 5

You are given three solutions containing three different salts — A, B and C — of *d*-block metals.

(a) **Write down the colour of each solution.** (3)

(b) **To a portion of solution A, add ammonia solution until in excess. Write down your observations.** (2)

(c) **To a second portion of solution X, add $1\,cm^3$ of dilute hydrochloric acid followed by $1\,cm^3$ of barium chloride solution. Write down your observations.** (1)

(d) **Identify X by writing its formula.** (1)

(e) To separate portions of solutions **B** and **C**, add aqueous sodium hydroxide until in excess. Write down your observations with **B** and with **C**. **(3)**

(f) To a second portion of **B**, add 5 drops of potassium iodide solution followed by 5 drops of starch solution. Write down your observations. **(2)**

(g) Identify the cations present in **B** and **C** by writing their formulae. **(2)**

Total: 14 marks

Test question 6

(1) Use a measuring cylinder to measure $30\,cm^3$ of $1.0\,mol\,dm^{-3}$ sulfuric acid and pour it into a beaker.

(2) Clean some magnesium ribbon with sand paper and cut it into 4 equal lengths between 1 and 2 cm.

(3) Measure the temperature of the acid. Then add one length of magnesium ribbon and start a clock.

(4) Stop the clock when the magnesium disappears.

(5) Measure out another $30\,cm^3$ of acid and warm it on a tripod and gauze until its temperature reaches just under 30°C.

(6) Remove from the heat, measure its temperature and add a piece of magnesium ribbon. Time how long it takes for all the magnesium to react.

(7) Repeat steps 1–6 at 40°C and 50°C.

(8) Record all your measurements in a table (see Table 1). **(2)**

(9) Copy and complete Table 2. **(4)**

Table 1

Volume of acid/cm³	Length of magnesium/cm	Temperature/°C	Time/s

Table 2

$\frac{1}{t}/s^{-1}$	$\log_{10}\left(\frac{1}{t}\right)$	Temperature, T/K	$\frac{1}{T}/K^{-1}$

You may assume that $\log_{10}\left(\dfrac{1}{t}\right)$ is proportional to $\dfrac{-E_a}{2.3RT}$, where t is the time, T is the temperature in kelvin and $R = 8.31\,\text{J}\,\text{K}^{-1}\text{mol}^{-1}$.

(a) Draw a graph of $\log_{10}\left(\dfrac{1}{t}\right)$ on the y-axis against $\dfrac{1}{T}$ on the x-axis. **(4)**

(b) Calculate the slope of the graph. Show on the graph how you did this. **(1)**

(c) Calculate the activation energy for this reaction. Give a sign and units with your answer. **(2)**

(d) Suggest *one* main source of inaccuracy in this experiment. **(1)**

Total: 14 marks

■ ■ ■

Answers to Practice examples and Test questions

Practice examples

Practice example ASB1

(a) Lilac flame ✓

(b) Drops of water condense in upper part of test tube ✓; the limewater goes milky ✓

(c) The solution stays clear and colourless ✓

(d) $KHCO_3$ ✓

Practice example ASB2

(a) Forms two layers ✓; the universal indicator goes green ✓

(b) Steamy fumes produced ✓; litmus paper goes red ✓

(c) The solution stays orange ✓

(d) The peak at P is caused by the O–H bond ✓ and the peak at Q by a C–H bond ✓

(e) ✓

Practice example A2B1

(a) A gives an orange precipitate ✓; with B there is no change ✓

(b) With A there is no change ✓; B produces steamy fumes ✓

(c) With A the solution stays orange ✓; B turns the orange solution green ✓

questions & answers

(d) As A is a carbonyl compound (test 1), but is not an aldehyde (not oxidised in **(c)**) and has 3 carbon atoms (given), it is propanone. It will have 1 peak ✓only (both CH_3 groups in the same environment) and this will not be split ✓ (because there are no hydrogen atoms on the adjacent carbon)

Practice example A2B2

(a) Fizzes ✓ and white solid formed/sodium disappears ✓

(b) Orange solution goes green ✓

(c) Pale yellow precipitate ✓

(d) P is butan-2-ol ($CH_3CH(OH)CH_2CH_3$) ✓

(e) Q has a C=O group ✓ (absorption at $1718\,cm^{-1}$) but no O–H group ✓ (no peak above $3000\,cm^{-1}$)

(f) The solution stays orange ✓

(g) Q is butanone ($CH_3COCH_2CH_3$) ✓

(h) With R, the solution fizzes ✓ and limewater turns cloudy ✓

(i) Fruity/gluey smell ✓

(j) Ester ✓

(k) It is CH_3CH_2COOH, propanoic acid

Practice example A2B3

(a) Green crystals ✓

(b) Water condenses in upper part of test tube ✓; the solid melts ✓; brown gas evolved ✓

(c) Blue precipitate ✓ which forms a dark blue solution with excess ✓

(d) Brown colour ✓ (with precipitate); on addition of starch a blue–black colour ✓

(e) $2Cu^{2+}(aq) + 4I^-(aq) \rightarrow 2CuI(s) + I_2(s)$ ✓   allow $I_2(aq)$

Practice example A2B4

(a) It is green ✓

(b) Green precipitate ✓ which turns brown ✓ on the sides of the test tube on standing

(c) White precipitate ✓

(d) Brown precipitate ✓

Practice example A2B5

(a) The solution is colourless ✓

(b) White precipitate ✓; which forms a colourless solution ✓ with excess

(c) White precipitate ✓; which forms a colourless solution ✓ with excess

(d) Cream precipitate ✓

(e) Solution goes red–brown ✓

Test questions

Test question 1

A is potassium sulfate; B is sodium hydrogencarbonate; C is a solution of magnesium chloride

(a) A gives a lilac ✓ flame colour; B gives a yellow ✓ colour (allow orange)

(b) White precipitate ✓

(c) K_2SO_4 ✓

(d) $Ba^{2+}(aq) + SO_4{}^{2-}(aq) \rightarrow BaSO_4(s)$ ✓

(e) Drops of water condense ✓ on upper part of test tube; limewater goes milky ✓

(f) Fizzes ✓ as B is added; limewater goes milky ✓

(g) $NaHCO_3$ ✓

(h) White precipitate ✓ which dissolves in ammonia solution ✓

(i) White precipitate ✓

(j) $MgCl_2$ ✓

Test question 2

P is propan-1-ol; Q is 2-bromo-2-methylpropane; R is cyclohexene

(a) Orange solution goes green ✓

(b) Steamy fumes ✓ evolved; blue litmus goes red ✓

(c) $CH_3CH_2CH_2OH$ ✓
(The reason is that it is an alcohol having an OH group and the *m/e* value of the molecular ion is 60. So there must be 3 carbon atoms (36), one OH (17) and 7 other hydrogen atoms making 60. The peak at 31 is 29 less than the molecular ion and results from the loss of a C_2H_5 group, so P must be $CH_3CH_2CH_2OH$, not $CH_3CH(OH)CH_3$.)

(d) Cream precipitate ✓; which stays on addition of dilute ammonia ✓

(e) $(CH_3)_3CBr$ or $(CH_3)_2CHCH_2Br$ ✓
(It is branched (the question says so). The two peaks of equal intensity of the molecular ion show that there are two isotopes with equal abundance. This must be Br and confirms the inference from the cream colour in **(d)**. $136 - 79$ (or $138 - 81$) leaves 57 which is made up of 4 carbon atoms (48) and 9 hydrogen atoms, So Q is 2-bromo-2-methylpropane or 1-bromo-2-methylpropane.)

(f) Burns with a smoky flame ✓

(g) Red–brown bromine goes colourless ✓ (goes clear would NOT score); two layers ✓ form

(h) C=C (alkene) group ✓; a high carbon to hydrogen ratio ✓

(i) Spectrum is of P ✓; broad peak at $3350\,cm^{-1}$ due to a hydrogen-bonded O–H group ✓

Test question 3

Z is hydrated ammonium iron(III) sulfate, $(NH_4)_2SO_4.Fe_2(SO_4)_3.24H_2O$
Samples of Z should be between 11.2 and 11.5 g in a weighing bottle
The sodium thiosulfate solution is $0.100\,mol\,dm^{-3}$

Scoring

- Table 1: all masses to at least 2 decimal places and subtraction correct ✓
- Table 2: all volumes recorded to 0.05 cm³ and all subtractions correct ✓
- Mean titre: correctly calculated ✓
- Accuracy mark: the expected titre = mass of Z × 2.074 cm³

Difference from expected titre	±0.3	±0.5	±0.7	±1.0	>1.0
Marks scored	4	3	2	1	0

Consistency mark: difference between outermost titres used to calculate the mean

Difference	0.2	0.3	0.5	>0.5
Marks scored	3	2	1	0

(a) Amount of sodium thiosulfate $= 0.100\,mol\,dm^{-3} \times \dfrac{\text{mean titre}}{1000}\,dm^3$ ✓

(b) Amount of $I_2 = \frac{1}{2} \times$ moles of sodium thiosulfate

Amount of Fe^{3+} in 25 cm³ $= 2 \times$ moles of I_2
Amount of Fe^{3+} in 250 cm³ $= 10 \times$ moles of Fe^{3+} in 25 cm³ ✓

(c) Mass of iron in 250 cm³ $= 55.8 \times$ moles of Fe^{3+} in 250 cm³ ✓

(d) % iron in Z $= \dfrac{\text{mass of iron in 250 cm}^3}{\text{mass of Z}} \times 100\%$ ✓

Test question 4

X is butan-2-ol, $CH_3CH(OH)CH_2CH_3$; Y is butanal, $CH_3CH_2CH_2CHO$; Z is butanone, $CH_3COCH_2CH_3$
[If there is any trouble getting hold of these substances, the propan- equivalents can be used]

(a) Has an O–H group ✓; but no C=O group ✓ (It is an alcohol)

(b) With acidified dichromate(VI):
X — orange solution goes green ✓ (It is a primary or secondary alcohol — see **(a)**)
Y — orange solution goes green ✓ (It is a primary or secondary alcohol or an aldehyde)
Z — solution stays orange ✓ (It probably is a tertiary alcohol or a ketone)

(c) With 2,4-dinitrophenylhydrazine solution:
X — no precipitate ✓ (It is not a carbonyl compound)
Y — yellow (or orange) precipitate is formed ✓ (It is an aldehyde — see **(b)**)
Z — yellow (or orange) precipitate is formed ✓ (It is a ketone — see **(b)**)

(d) In the iodoform test:

X — pale yellow precipitate ✓ (It contains $CH_3CH(OH)$ group and so is butan-2-ol)

Y — no precipitate ✓ (It is butanal; it cannot be methylpropanal because that has a branched chain — see stem of question)

Z — pale yellow precipitate ✓ (It contains a CH_3CO group, so is butanone)

(e) X is butan-2-ol, $CH_3CH(OH)CH_2CH_3$ ✓; Y is butanal, $CH_3CH_2CH_2CHO$ ✓; Z is butanone, $CH_3COCH_2CH_3$ ✓

Test question 5

A is a solution of nickel(II) sulfate; B is a solution of iron(III) chloride; C is a solution of chromium(III) sulfate

(a) Solution A is green ✓; B is yellow–brown ✓; C is green ✓

(b) Green precipitate ✓ formed; forms a pale blue solution with excess ✓

(c) White precipitate ✓ formed (note that it is a white precipitate in a green solution)

(d) $NiSO_4$ ✓

(e) B — red–brown precipitate which stays with excess ✓

C — green precipitate ✓; forms a green solution with excess ✓

(f) B gives a red–brown solution ✓; goes blue–black when starch is added ✓

(g) In B is Fe^{3+} ✓; in C is Cr^{3+} ✓

Test question 6

Table 1: All volumes ✓ and times ✓ recorded

Table 2: Correct calculation of: $\frac{1}{t}$ ✓; $\log_{10}\frac{1}{t}$ ✓; kelvin temperature ✓; $\frac{1}{T}$ ✓

(a) Graph: sensible scale and axes labelled ✓; points properly plotted and best-fit straight line drawn ✓

If all points close to the line ✓✓; if only one significantly off the line ✓

(b) Slope calculated properly ✓

(c) Activation energy $= (-\text{slope} \times 2.3 \times 8.31) \div 1000 \, \text{kJ mol}^{-1}$: value ✓; sign and correct unit ✓

(d) Either the temperature was not constant during each experiment; or the pieces of magnesium were not of exactly the same length ✓

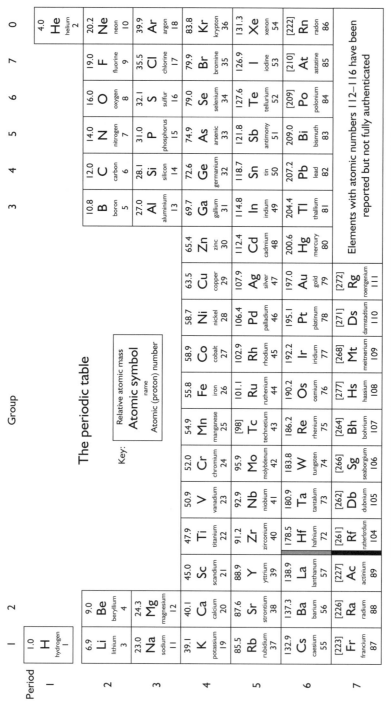

The periodic table

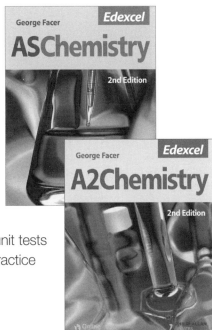